DATE			

Earthquake Country

By Robert Iacopi

Foreword by Dr. Charles F. Richter

LANE BOOKS • MENLO PARK, CALIFORNIA

Foreword

This is the third edition of a significant work which first appeared in 1964. The present form has been revised to bring maps and descriptive text up to date as nearly as practicable; it includes new illustrations, particularly for the very important San Fernando earthquake of February 9, 1971.

The San Fernando earthquake was one of the most important in the history of modern seismology. This is due to its location, to the numerous qualified investigators, and to the exceptional number of valuable instrumental recordings. Among significant results were: (1) detailed study of large-scale thrusting of a crustal block; (2) faulting along lines previously mapped but not then recognized as active; (3) deep fracturing along lines transverse to the surface geologic trends; (4) proof, from instrumental recordings, of accelerations in earth vibration comparable with that of gravity, subjecting buildings and other structures to forces comparable with their weights; (5) good performance of modern high-rise buildings in areas of moderate shaking; (6) failure of modern structures in the area of heaviest shaking.

The San Andreas Fault is the largest and most important feature of California considered as earthquake country; naturally it takes up the lion's share of space in the book. However, the reader will find much information here about other active faults—not only those of the larger San Andreas fault system, but others not directly connected. Earthquakes associated with all these, including the newly prominent San Fernando fault system, are discussed at appropriate length.

This wider perspective is wholesome, and in the public interest. Many persons still retain the idea that all California earthquakes are on or related to the San Andreas Fault. Correspondents write: "I understand that my residence is far from the San Andreas Fault; am I not then right in that there is no earthquake danger here?" No locality in California is exempt from earthquake risk; the only places of relatively low risk are areas of unfractured rock high up in the Sierra Nevada.

Mr. Iacopi has written a factually sober—and sobering—book. Those looking for exciting and sensational material will be disappointed. The serious-minded citizen will accept the publication as a public service.

Charles F. Richter
Pasadena, June 20, 1971

ACKNOWLEDGMENTS

Many geologists, seismologists and engineers made significant contributions of time and research materials to the preparation of this book. I am particularly indebted to these contributors:

Dr. Clarence R. Allen and Dr. Charles F. Richter of the California Institute of Technology; Karl V. Steinbrugge of the Pacific Fire Rating Bureau; the staffs of the U.S. Geological Survey, the U.S. Coast and Geodetic Survey, the California Division of Mines and Geology, and the University of California Seismographic Station.

ROBERT IACOPI

Colored lines used to show faults on maps and photos in this book are based on published geologic sources and have been checked by experts; however, these are intended to show only general path of faults and should not be interpreted literally

Third Printing May 1973

Contents

California Is Earthquake Country

EARTHQUAKES ARE A PART OF CALIFORNIA'S
HERITAGE, AND WE ALL MUST LEARN TO LIVE
WITH THEM. BUT THE DANGERS INVOLVED ARE
MORE A RESULT OF MAN'S IGNORANCE THAN OF
NATURE'S DESTRUCTIVE FORCE.

DESPITE ITS GREAT VARIETY of natural wonders, California is best known to many outsiders for just one thing — earthquakes. California is earthquake country, and the many reports and stories on its rumblings and shakings have made indelible impressions throughout the world.

Though sometimes exaggerated, California's position in the world of earthquakes is indeed a significant one. It is part of the circum-Pacific seismic belt that is responsible for about 80 per cent of the world's earthquakes. All parts of this belt are annually jolted by countless numbers of major and minor shocks, and California is no exception. Other spots around the Pacific Basin—such as Japan and the Aleutian Islands —may have more earthquakes than California, but the West Coast state still is hit by thousands of shocks every year, some 500 of which are large enough to be felt by many people. Earthquakes of destructive magnitude have occurred in California on an average of once a year for the past 50 years, and few earthquakes in the world received as much publicity as did the 1906 California quake.

Other parts of the United States have also been struck by earthquakes; three of the largest on record were at New Madrid, Missouri (1811), Charleston, South Carolina (1886), and Anchorage, Alaska (1964). But California has a clear claim on earthquake frequency in this country, and its reputation as earthquake country is kept alive by the many minor shocks that strike annually, and by the generous amount of newspaper space that is afforded them.

But bare statistics do not tell the whole story. Any Californian who knows even a few facts about the state's seismic history realizes that "earthquake" can mean many things. Most often, it refers to minor quivers of the earth that people can feel but that rarely do more than rattle windows, crack a little plaster, or occasionally knock an older home off its foundation.

Earthquakes of this size are so common in some coastal areas that anyone who follows the newspapers regularly may soon fancy himself an expert on the subject. After every shake, magnitude numbers are repeated and compared like so many hat sizes, and the most popular guests at dinner are those who can relate exciting personal experiences or can quickly compare the most recent shaking with "the last big

California's susceptibility to devastating earthquakes was dramatically demonstrated on February 9, 1971, when the southern half of the state was jolted by a tremor centered in San Fernando. Sixty-four people died, including 44 in the collapse of the Veterans Administration Hospital at Sylmar (shown above).

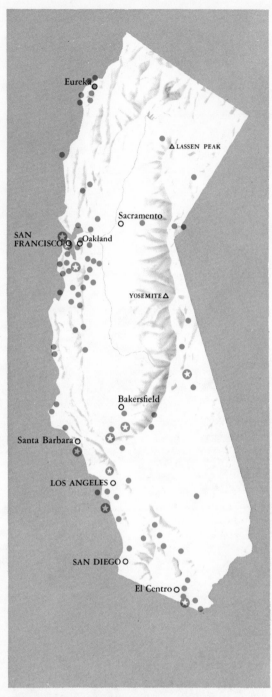

This map shows most of California's historic quakes (stars indicate the shocks that resulted in greatest intensities). A continuation of this same general earthquake pattern is expected in the future.

one." Earthquakes provide a few thrills and plenty of conversation, but most of them offer little to worry about.

The fact that California averages one earthquake of destructive magnitude per year does not mean that a *city* of some size is annually flattened. On the contrary, a great percentage of these larger earthquakes are centered far enough from population centers that the great shock intensities close to the source do nothing more than shake some sand dunes or frighten a few cattle. The larger cities—far in the distance—feel nothing more than the last, dying shudders.

However, for the resident of Santa Rosa in 1906, Santa Barbara in 1925, Long Beach in 1933, or San Fernando in 1971, "earthquake" means something else again. It is a terrible and devastating force that can send violent convulsions through the earth and shake buildings to the ground in seconds.

A gentle swaying of a house and a few rattling dishes are one thing, but when the neighborhood school is thrown down in a heap of rubble and your garage roof collapses a few feet from the back door, "earthquake" takes on new meaning.

It is unfortunate that many of California's smaller quakes are magnified into near-disasters by hasty newspaper reports and the wild spreading of rumor. The slightest tremor can cause a panic, and even the most reasonable of men have been known, during times of panic, to ascribe earthquake effects to the most illogical sources. Since few people know a great deal about the origin and motion of earthquakes, anyone with even the barest knowledge may soon be converted into an authority. Those who can rattle off half-remembered stories and isolated statistics, and mention the "San Andreas Fault" in the same sentence, are often the worst culprits.

Compounding the problem is the uncritical receptiveness of many earthquake victims. Too often they are frightened people, eager to grasp at random bits of information. Without the proper background and context, new-found knowledge may be quickly distorted and passed on in garbled form. The files of seismological stations are filled with letters of preposterous, unbelievable stories that undoubtedly were born during those hours of crisis.

Despite the public attention concentrated on recent earthquakes, California has had only three "monster" earthquakes during recorded history—1857, in Southern California; 1872, in Owens Valley; and 1906, along the Northern California coast. During the past 50 years, only the 1952 Kern County, the 1968 Borrego Mountain, and the 1971 San Fer-

The 1933 Long Beach earthquake focused attention on the inadequacies of California's building codes. Many school buildings, including Jefferson Junior High School, offered little resistance to shock.

The Hayward earthquake of 1868 ranks among California's greatest shocks. The ground was ruptured for 20 miles, and damage was heavy throughout the Bay Area; old engraving is of a mill in Hayward.

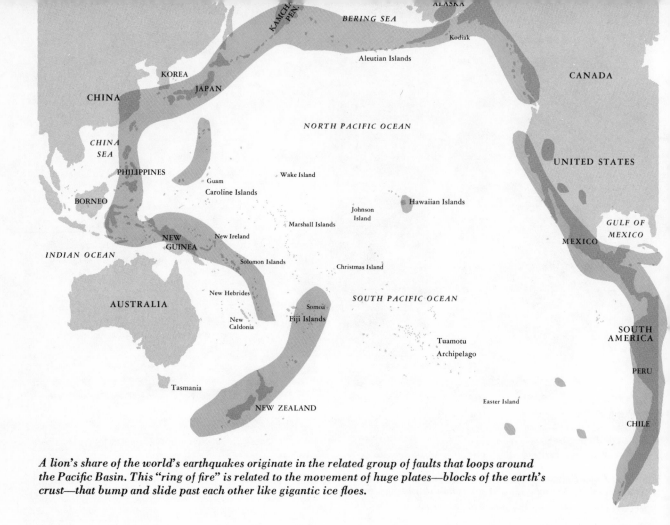

A lion's share of the world's earthquakes originate in the related group of faults that loops around the Pacific Basin. This "ring of fire" is related to the movement of huge plates—blocks of the earth's crust—that bump and slide past each other like gigantic ice floes.

nando quakes have even approached these in size. Damage and loss of life, on the other hand, have not been proportionate. The quakes of 1925, 1933, and 1971 (all in southern California) are relatively moderate, but they wreaked havoc in densely populated areas. Because of the highly publicized results, these three are often thought of as maxi-quakes, when in fact the loss of life and property damage were far out of line with the actual size of the earth movement.

The history of California earthquakes started in 1769, when the expedition of Gaspar de Portola was violently shaken by a large earthquake while they were camped on the Santa Ana River near the present town of Olive. The heavy shaking reportedly threw the river out of its channel, and many men and horses were knocked to the ground. Portola named the river "The River of the Sweetest Name of Jesus of the Earthquakes."

Because of the slow rate of development in the state during the next century, earthquake reports were practically non-existent until the 1850's, when the population upsurge also brought increased interest in California's geology. Around 1900, scientific inquiries

became more numerous. Then the 1906 disaster focused full attention on the earthquake problem in California and started the development of modern seismology.

What we know as earthquakes are only a small part of a greater geologic process that is constantly taking place in the earth's crust. When the earth shifts along lines of weakness known as faults, the violent movement of great rock masses creates shock waves.

We can learn a great deal about the nature of earthquakes by studying the faults along which they originate and by observing the physical features that have been created by both ancient and recent earth movements. California's San Andreas Fault is by far the most conspicuous in the state and is so well delineated in many areas that it can be easily explored by the layman.

Perhaps the most significant lesson to be learned from the study of geology is that an earthquake is not capricious. Anyone who has lived through a large shake will surely remember some oddity that made the earthquake seem to have a mind of its own. One house may have stood firm while its neighbor col-

The earthquake of 1906, and the fires that followed, resulted in the greatest seismic disaster in California history. Every city along the San Andreas Fault from Hollister north to Fort Bragg was badly shaken. Central tower on San Francisco's Hall of Justice was badly damaged by the initial shock waves, and the fire that followed gutted the lower levels.

lapsed; or persons living close to the fault may not even have realized that an earthquake was going on, while others living 15 or 20 miles from the earthquake center were knocked down. There is nothing mysterious about these inequities. Each effect is based on a sound geologic principle that can be applied to all earthquakes and therefore can be used to predict the probable results of the next major shock.

After every large tremor, the various governmental and private organizations that are concerned with earthquakes are flooded with questions from the public: Why do we have earthquakes? Why is one bigger than another? How strong is an earthquake? What can be done about all this? But these are the questions that should be asked not *after* an earthquake, but *before*. Investigations of causes and probable effects have progressed to such an extent that every Californian can save himself a great deal of worry and

grief—and perhaps thousands of dollars—by understanding the facts and learning to live with them.

The shaky Californian who has just been through a heavy jolt and looks for some way to escape future danger will be acutely disappointed. There is no way to escape earthquakes and still live in California. It is true that there are some regions that are less susceptible to earthquakes, but they tend to be in the more remote areas away from the centers of industry and commerce. The most practical attitude is to admit that California is earthquake country and that it will continue to be for countless centuries. And while there is no way of eliminating causes, there is much that can be done about effects. Only one in 10,000 California earthquakes ever does any great damage, but you should always be ready for the day when that almost casual shaking under your feet suddenly becomes the greatest earthquake of a decade.

Great Faults Slice through the State

EARLY INDIANS THOUGHT THAT A TORTOISE SUPPORTED THE EARTH; WHEN THE BEAST TOOK A STEP, THE EARTH TREMBLED. TODAY WE KNOW THAT EARTHQUAKES ARE ONLY PART OF A GREAT GEOLOGIC PROCESS.

CALIFORNIA IS BROKEN into a series of crustal blocks that are separated by faults—great fractures that form lines of weakness in the masses of rock at the earth's surface. Due to pressures that build up below the surface, these blocks are elevated, tilted, folded, and depressed along the fault lines until they resemble the pattern of an old brick street. And every time two of these blocks suddenly shift past each other along a fault, California has an earthquake.

Part of the mystery surrounding faults and their movements has been explained by the theory of continental drift, or "plate tectonics." This theory is commonly accepted by most scientists today.

According to continental drift, all the earth's land area once was a single mass. This mass broke apart, and the pieces have been "drifting" around the globe ever since. The reason behind the movement is the constant creation and destruction of huge masses of material called plates. These rigid masses float on a hot, viscous substance deep within the earth. The continents are lighter than the plates and ride along on top of them—moving as the plates move.

New plates are formed at weak spots in the earth's crust—"spreading centers"—generally thought to be high ridges in the mid-ocean floors. Hot material from the earth's mantle oozes up along these ridges, solidifies, and then moves laterally away from the ridges at approximately 2 inches a year. As the new material is created, the energy pushes the existing crust outward and the old edges of the plates are shoved down into ocean trenches where they remelt and start the process all over again.

Wherever plates meet or an old plate plunges into a trough, there is a high seismicity. Off the west coast of South America, a huge plate is being shoved into a deep trench. As it moves, it pushes up the Andes mountains and causes tremendous earthquakes.

Along the west coast of North America, one plate is thought to comprise the whole northern Pacific Ocean and a slice of far western California. Another plate includes the rest of North America and the western half of the Atlantic Ocean. They are divided by the San Andreas Fault. As the Pacific plate moves northwesterly toward the Aleutian trench, and the North American plate heads southeasterly, they grind

The great San Andreas Fault, dominant part of California's extensive fault system, has carved a conspicuous path through coastal mountains and southern desert. On the San Francisco Peninsula, the characteristic rift valley is partly submerged under Crystal Springs and San Andreas Lakes.

Clarence R. Allen

Around the rim of the Pacific basin are active faults—such as the Atacama Fault in Chile (above) and the ones pictured on the opposite page—all a part of a great pan-Pacific chain that includes California's fault system. Chile's Atacama Fault resembles California's San Andreas Fault in nature of horizontal movement, straight line of fracture, and great number of offset streams.

against each other along the fault—and California has earthquakes.

The ocean floor ridge between these two plates—and the source of the San Andreas Fault—is called the East Pacific Rise.

These theories have thrown new light on the origin, frequency, strength, and perhaps even prediction of earthquakes in California and the rest of the world.

CALIFORNIA'S FAULT SYSTEM

California is interlaced with hundreds of faults, the most important of which are shown on the map on the facing page. Those singled out for special attention

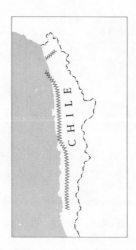

The Atacama Fault of Chile runs through mountainous terrain in a course that parallels the coast, as does the California system.

The Philippine Fault is hidden by a thick jungle cover. But recent geologic investigations have shown that it, too, is similar to the San Andreas Fault in many respects.

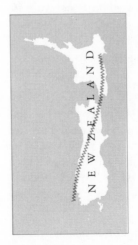

The Alpine Fault of New Zealand appears as a great scar across the landscape. River courses here tend to follow straight fault lines, as is the case along the San Andreas zone.

Movements along the Garlock Fault have upended and deformed the rocks that form the south end of the Slate Range. Flat, undisturbed surface at right is the alluvial outwash from the range. In the background is Brown Valley and the southern tip of the Panamint Valley.

have either made significant contributions to the visible land forms of the state, or have caused major earthquakes during recorded history.

While all of the faults in California may be considered as parts of the same general system, there are distinct variations in direction and movement. The San Andreas and related faults are arranged in a southeast-northwest direction, while another group—dominated by the Garlock Fault—runs predominantly southwest-northeast.

The effect of the faults on California becomes readily apparent when they are plotted on a topographic map of the state. The Mojave Desert, for example, is clearly limited on the south and west by great faults, and the development of the San Francisco Bay area has been substantially influenced by movements along branches of the San Andreas system.

Each of the major faults shown on the map consists of several interwoven branches that together make up a definable fault zone. Almost all of the surface details that we can trace today are located along the branches that have experienced the most recent movements. In most cases, the next shift within any given fault zone can be expected to take place along the same active branch where the last movement occurred.

SIGNIFICANT FAULTS IN CALIFORNIA

SAN ANDREAS FAULT is the most publicized rift in California. It is by far the longest in the state, and it annually produces dozens of earthquakes. But despite its importance, the role of the San Andreas is often exaggerated; it is frequently blamed for every earthquake in California, and many people believe that once they move away from this great fault, they no longer need to fear earthquakes—a dangerous fallacy.

HAYWARD FAULT, despite its distinctive name, is really a branch of the San Andreas zone. The fault has played a significant role in the geologic development of the San Francisco Bay area, and it has also given birth to several large tremors.

SIERRA NEVADA FAULT movements have created the magnificent escarpment that forms the eastern edge of the Sierra. The Owens Valley branch of the system was responsible for the 1872 quake—the largest in California's recorded history.

WHITE WOLF FAULT is a short, relatively insignificant fault that unexpectedly moved in 1952 to cause a major quake in the Arvin-Tehachapi area.

GARLOCK FAULT is the second largest fault in the state, and has made several contributions to the landscape, including the mountain ranges that form the northern edge of the Mojave Desert. Strangely enough, there has not been a single great earthquake during recorded history that can be blamed on this huge fracture.

SANTA YNEZ FAULT is the largest of a group of related breaks that form a large seismic area around the Santa Barbara channel. The most spectacular earthquake to originate in this region was the 1925 Santa Barbara shock.

SAN FERNANDO FAULT, very similar to White Wolf Fault, was responsible for the devastating 1971 San Fernando earthquake. This fault was relatively inactive in preceding years, and the 1971 break came as a surprise.

NEWPORT-INGLEWOOD FAULT was unknown until 1920, when a small earthquake told seismologists that there was an active break along the coast. If there were any doubts about activity, they were dispelled in 1933, when the disastrous Long Beach quake rattled the coast.

SAN JACINTO FAULT is part of the San Andreas zone—indeed, it may be the most active branch. It has been the source of many important quakes, and the land forms along its route give mute testimony to its long-term significance on the state's topography.

IMPERIAL FAULT is another branch of the San Andreas zone. Exact route of the fault was not exposed until the 1940 earthquake which ruptured the surface on both sides of the U.S.-Mexico border.

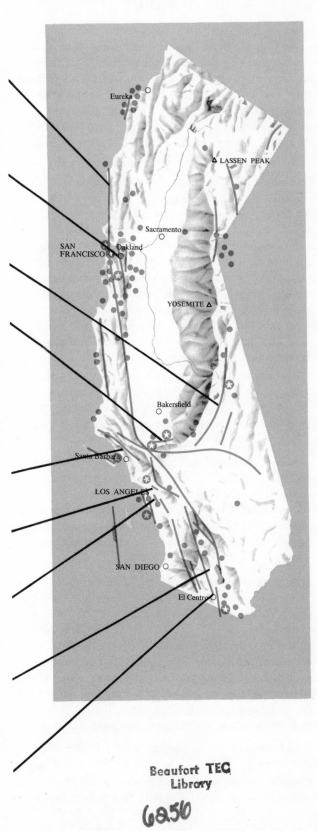

THE SAN ANDREAS FAULT

The most dominant fault in California today is, by far, the San Andreas Fault, a giant shear zone that extends some 650 miles through Southern California and along the Coast Range section of Central California. This huge fissure has been instrumental in the development of the topography all along its course, and has caused two of the greatest earthquakes in California's history (1857 and 1906) plus a number of others that have excited public interest.

Because of its immense size and importance to California, the San Andreas Fault has been carefully studied ever since its existence was recognized. The first notions that a huge fault did actually cut California came in 1893, when the noted geologist Andrew Lawson took a steamer trip from San Diego to San Francisco, stopping off to examine the coastal development at several places. He noted that San Benito, Santa Clara, and the San Francisco Bay valleys were remarkably linear, and that the general uplift of the continental margin was marked by tilting and movements that "may yet be active."

The fault received its name two years later, when Lawson pointed out that the fault features were best exposed and typified by the straight valley on the San Francisco Peninsula that was occupied in part by San Andreas Lake.

The San Andreas Fault became known around the world after the 1906 earthquake. This event was responsible for some 700 deaths and millions of dollars of damage, and it opened the door to a new line of scientific thought that forms the basis of many of today's basic concepts about faults and earthquakes.

Like all major faults, the San Andreas is not a single break in the rock, but is a wide zone, made up of several lines of activity that are roughly parallel. It is not of a single age, but includes the remnants of ancient faults that have been quiet for countless centuries and other active breaks that form the lines of most recent activity within the zone. The discontinuous movement along the fault has given rise to a confused surface appearance consisting of old features that have been heavily eroded, plus the fresher results of movements of the past few thousand years.

Most California earthquakes originate at points about 10 miles deep, but the San Andreas Fault undoubtedly extends down through the earth's crust (a distance of some 20 to 30 miles) to the East Pacific

Rise, one of the spreading centers associated with continental drift and plate tectonics (see page 10).

The San Andreas Fault is traceable as a continuous belt from San Gorgonio Pass in the south to Point Arena on the Mendocino County coast. South of San Gorgonio Pass, however, the fault is a wide zone, and several parallel branches together make up the San Andreas system.

South into Mexico, geologic evidence and the location of earthquakes have long indicated an almost certain continuation of the fault into the Gulf of California. Then, in 1971, a research team from the University of Southern California came up with enough new evidence to establish that the southern end of the San Andreas definitely is in the Gulf, opposite the town of San Felipe. The East Pacific Rise appears to enter the Gulf of California, and the San Andreas Fault is a major break in it.

The fault's northern end has been firmly fixed only in recent years. During the 1906 earthquake, there was a surface rupture at Shelter Cove on the Humboldt County coast, leading some geologists to believe that this was a northward extension of the San Andreas Fault. It was even thought that the main fault line might even extend in a straight line toward Alaska. But new evidence indicates that the fault line ends at the Juan de Fuca plate, a small piece of a much larger plate that is plunging into an ocean trench off the Oregon and Washington coasts and seems to be instrumental in the further development of the Cascade ranges.

In light of this new finding, the Shelter Cove rupture in 1906 has now been blamed on a separate fault that was triggered by the main break along the San Andreas.

Within California, the San Andreas varies considerably in width. In places, it may be less than 100 yards wide and made up of but a few entangled lines of rupture. In most sections, however, it is several hundred yards to a mile or more in width and is interlaced with any number of sub-parallel fault lines. Its actual edges are indefinite because of the many old lines of activity that are now hidden under recent gravel deposits or alluvium, and because of the landsliding that has covered several miles at a stretch. South of San Gorgonio Pass in Southern California, the San Andreas zone is so wide that almost the entire width of the state falls under the spell of this great cleft and its many branches.

Intermittent movements, plus erosion along the crushed rocks within the fault zone, have created a distinct valley along the Cucapa fault, a branch of the San Andreas system in Baja California.

Fault Action Triggers Earthquakes

CALIFORNIA IS SLOWLY SHIFTING BENEATH OUR FEET, ACCUMULATING STRAIN THAT NOT EVEN THE EARTH CAN BEAR. EVENTUALLY, AN ADJUSTMENT MUST BE MADE—AND AN EARTHQUAKE IS BORN.

FAULT MOVEMENT IS THE RESULT OF elastic rebound—the slow build-up and sudden release of strain within masses of rock.

The forces responsible for mountain building exert great pressures at the surface of the earth and cause crustal blocks to move relative to one another. For example, surveys on the San Andreas Fault in Northern California have shown that the area west of the fault is gradually drifting past the rocks east of the fault at a rate of about two inches a year, or 15 to 20 feet in a century. This very slow movement constantly introduces strain into the rocks on each side of the fault.

In opposition to the steady build-up of strain is the basic elastic strength of the rocks along the fault. The rocks may become deformed and their positions distorted, but they tend to hold their basic positions and the fault remains "glued" together by friction.

Eventually, the strain build-up becomes so great that it overcomes the strength of the rocks. Then the earth moves, and masses of rock on opposite sides of the fault scrape past each other. It is the jarring "fling" of the rocks past each other that creates earthquake waves.

Usually, the resistance of rocks along a fault is relatively small, and the earth moves after only a slight build-up of elastic strain. The adjustments therefore are small, and the earthquakes are minor. But when the strength of the rocks is very great, or if the fault is "glued" tightly, then pressures may build up for a hundred years or more and assume massive proportions. When the stress finally becomes great enough to overcome the tight bond—and it most certainly will—the pressure is released like the uncoiling of a great spring, and the action of the rocks sliding past each other creates a major quake. The earthquakes of 1857, 1872, or 1906 were of such proportions.

While fault movement certainly starts at a single point, strain can be relieved along many miles of the fault line. The total displacement may not be more than 10 or 20 feet, but surface movements of this magnitude may take place at several different spots.

Fault movements can be horizontal, vertical, or a combination of the two. The San Andreas Fault apparently has gone through long periods of both vertical and horizontal shifting. The great, eroded

THEORY OF EARTHQUAKE MOVEMENT

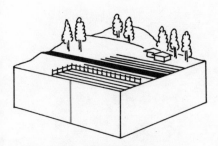

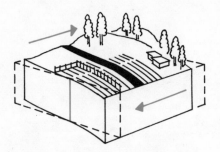

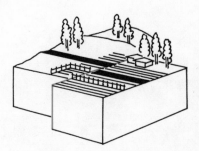

According to the Elastic Rebound Theory, a fault is incapable of movement until strain has built up in the rocks on either side. This strain is accumulated by the gradual shifting of the earth's

crust (at a rate of about two inches a year along the San Andreas Fault). Rocks become distorted but hold their original positions. When the accumulated stress finally overcomes the resistance of

the rocks, the earth snaps back into an unstrained position. The "fling" of the rocks past each other creates the shock waves we know as earthquakes.

escarpments along the fault in Southern California —including the steep faces of the San Jacinto and San Bernardino mountains—clearly indicate that during ancient geologic times vertical displacements totaled some 10,000 to 15,000 feet. The 1906 earthquake, on the other hand, was the result of a horizontal fault movement that reached 20 feet at the surface. Vertical movements were rare, and never exceeded 3 or 4 feet.

Measurements of fault movements are always stated in relative terms. When fault movement is vertical, there can be either an uplift of the earth on one side of the fault, or a depression of the earth on the opposite side. In some cases, both actions may occur simultaneously, or an uplift may occur in one movement and a depression in the next. Horizontal mea-

surements need the same careful wording. Along the San Andreas Fault, the coastal (western) portion of California is said to move north in relation to the continental mass that lies east of the fault, though there is no way of knowing which side actually does the moving.

Horizontal movement may be either right- or left-lateral. The San Andreas is an example of a right-lateral fault. If you stand on either side of the fault line and look across it, the movement of the opposite side is always to the right. While there may be evidence of minor reversals of trend in any given movement, the long-term direction remains the same. All of the faults related to the San Andreas system have this same right-lateral movement. Other California faults, including the Garlock, are left-lateral.

HOW A FAULT MOVES

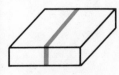

Fault movements must always be expressed in relative terms, since it is impossible to tell which side actually does the moving.

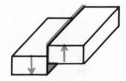

Vertical displacement can be created by either an upthrust of one block or a downthrow of the other. Most California earthquakes during historic times have been characterized by vertical shifts of only a few feet.

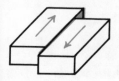

Horizontal movement can be either left- or right-lateral (shown here). The San Andreas and related faults all move right-laterally. The greatest single shift was 20 feet during the 1906 earthquake.

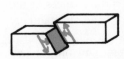

In thrust faulting, one block moves up and over another. This type of movement characterizes the White Wolf and San Fernando fault systems, which caused the 1952 and 1971 earthquakes.

Twenty-foot scarp resulting from a single fault movement in 1872 still stands out from the floor of Owens Valley. The movement that created this new feature also produced the greatest earthquake in California history. The Alabama Hills and massive eroded escarpment of the Sierra Nevada are visible in the background.

THE FAULT AT WORK—FAST CHANGE

When earth pressures are suddenly released in a single, violent fault movement, the surface may be broken along the fault line above the points of maximum dislocation. In less than a minute, the surface movement can cause some spectacular landscape changes. Vertical fault movements, with their shifts in elevation, create the most easily recognized changes. But a great earthquake that is characterized by predominantly horizontal movement can also leave its mark, as the photographs on these pages indicate.

A single earthquake may leave an impressive trace which is recognizable for a few decades or perhaps a century or two, but these changes are minute when compared with the long-term work of great faults such as the San Andreas. Most of the surface traces of the 1857 and 1906 earthquakes have been substantially modified by erosion or covered by landslides and other recent changes in surface contours, and only a few features are still recognizable. The scarps and cracks formed by the 1952 Kern County earthquakes on the White Wolf Fault were scarcely visible ten years later.

Rows in this orange grove (now removed) were offset 10 feet by the 1940 Imperial Valley earthquake.

Robert E. Wallace

Slow creep at the winery near Hollister has broken the concrete drainage ditch located south of the main building. Winery walls also are cracked.

This is one of a dozen places in Hollister where slow creep along the fault line is distorting curbs, sidewalks, gutters, and driveways.

But nature is patient. The accumulation of hundreds or thousands of movements, over millions of years, eventually bring indelible changes in the face of the land.

THE FAULT AT WORK—SLOW CREEP

In addition to the sudden fault movements that produce earthquakes, there is also another type of slower, gradual movement that may be equally important. During the last half-century, triangulation measurements across major fault lines have indicated a few isolated instances of gradual slippage, but they have been almost impossible to accurately trace and measure. In 1956, quite by accident, a remarkable example of such slow creep was discovered at a winery south of Hollister.

During a routine investigation into some unexplainable damage on the winery property (now owned by Almaden Vineyards), it was found that reinforced concrete walls and floor slabs of one warehouse were being fractured. Since the building is located within the San Andreas Fault zone, seismologists immediately looked into the possibility of fault movements and discovered that an active branch of the fault zone runs right underneath the building and that the two sides of this fault line are steadily moving past each other at an average rate of one-half inch per year. The movement is gradually pulling the building apart.

In accord with the right-lateral nature of the San Andreas Fault, the west side of the building is steadily moving north in relation to the east side. Winery employees regularly patch the floor cracks and shore up the sagging walls. The present buildings were constructed in 1939 (the former building was so weakened by the fault movement that it eventually collapsed), so the accumulated movement has not been sufficient to destroy the building—as yet. Measurements made since 1956 show that the fault movement continues, and there is no reason to believe that it will stop in the near future.

The creep is not entirely gradual. Recorders show that movement generally takes place in week long periods, though the movements may be separated by weeks or months. "Instant creep" of one-eighth inch took place during an earthquake in 1960, and the frequency of other earthquakes may have a bearing on the periods of activity. An earthquake on June 24, 1939, caused adobe walls to pull away from side walls, girders to separate from brickwork, and new cracks to open in the ground around the building. A severe jolt in 1961 also shook the buildings substantially and caused minor damage.

The slow creep at the winery is under constant study to determine its significance to the area's earthquake activity—or lack of it. Other areas of slow creep have been discovered in recent years—particularly in Hollister—and the whole subject has assumed new importance as scientists try to analyze the movements. It is interesting to note that slow creep is restricted to certain areas along the fault, and that other areas are completely free of such motion. For example, the

Elizabeth Lake tunnel for the Owens Valley aqueduct, which crosses the San Andreas Fault at right angles, was completed in 1911, but it has not been appreciably offset since that time.

Correct analysis of slow creep and its causes may yet be one of the most important steps in the understanding of quakes, and perhaps in their prediction.

BAJA CALIFORNIA

The most striking theory regarding accumulated horizontal displacement along a fault system concerns the origin of Baja California and the Gulf that separates it from the Mexican mainland.

The recent research in plate tectonics (see page 10) and the location of the southern tip of the San Andreas Fault (see page 16) lend new credence to the long-accepted idea that the Gulf of California was created by movements along the fault and continues to be influenced by such movements.

Southeast of Baja, the straight coastline of the Mexican mainland is broken by an abrupt southwestern swing of some 100 miles near Jalisco. The shape of this indentation roughly approximates the shape of the southern tip of the Baja peninsula. Geologic features suggest that Baja may have once lain against this coastline as part of the mainland, but that horizontal movement along the San Andreas Fault caused the peninsula to break away and move northwest, thereby forming the unusual deep water gap of the Gulf of California. The total horizontal movement that would be required for this shift is only 250 miles, which is certainly reasonable in light of the estimated movements within California.

Evidently, the Gulf is being widened at the rate of about two inches a year—the same rate that characterizes stress buildup along the San Andreas Fault in California.

THE SIERRA NEVADA

The most spectacular example of accumulated vertical movement along a fault line in California is the eastern escarpment of the Sierra Nevada.

The Sierra Nevada is a great fault block that has been slowly raised and tilted westward over several million years. Its steep eastern face clearly marks the line of the boundary fault.

This scarp is highest and most clearly defined between Olancha, at the Southern end of Owens Valley,

and Big Pine, where the steep face of the range is not broken by any foothills. Ridge spurs that have been truncated by the faulting may be seen south of Owens Lake and southwest of Independence.

West of Independence, the contour of the Sierra scarp changes, with the addition of foothills that have not been depressed as far as Owens Valley and have not been lifted as far as the main block. But as far north as Birch Mountain, the eastern face of the range presents unmistakable proof that it has been developed by repeated dislocations along the fault system.

HORIZONTAL MOVEMENT ALONG THE SAN ANDREAS FAULT

Since the San Andreas is an active fault of great size, a study of the movement that has taken place along it provides a clue to the effect of California's fault system on the geologic development of the state.

Recent activity along the San Andreas has been predominantly horizontal. This type of movement is measured in terms of displacement—the amount of separation between two rock types that once faced each other across the fault line. In theory, the object in making such measurement is to locate a certain rock formation on one side of a fault, then find that same rock on the other side and measure the displacement. By comparing the displacement with the age of the rock, the amount of movement during various geologic epochs can be accurately computed. At no point along the San Andreas Fault do old rock types on opposite sides of the fault match, so there is ample opportunity to investigate possible horizontal movement.

This sounds much simpler than it is in practice, since finding matching rocks on both sides of a fault is not always possible. The very faulting action that causes the displacement may also uplift the formations to where erosion can wear them away, or depress the rocks to where younger layers of gravel and rock can bury them from sight. Then, too, it is not often possible to positively identify rock on one side as being an exact match of a similar deposit across the fault line.

The map on the opposite page shows seven instances where matching rock types may indicate displacement along the San Andreas Fault. The ideas were advanced in 1953 by geologists Mason Hill and T. W. Dibblee, Jr. It should be noted that not all geologists concur in the interpretation of the evidence, particularly in the case of older, greater movements.

HORIZONTAL MOVEMENT ALONG THE SAN ANDREAS FAULT

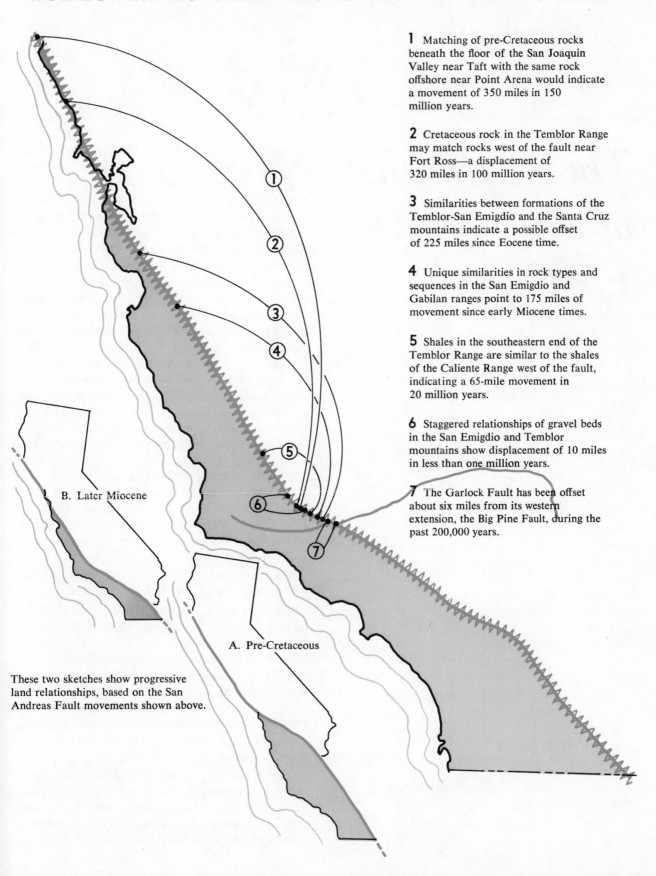

1 Matching of pre-Cretaceous rocks beneath the floor of the San Joaquin Valley near Taft with the same rock offshore near Point Arena would indicate a movement of 350 miles in 150 million years.

2 Cretaceous rock in the Temblor Range may match rocks west of the fault near Fort Ross—a displacement of 320 miles in 100 million years.

3 Similarities between formations of the Temblor-San Emigdio and the Santa Cruz mountains indicate a possible offset of 225 miles since Eocene time.

4 Unique similarities in rock types and sequences in the San Emigdio and Gabilan ranges point to 175 miles of movement since early Miocene times.

5 Shales in the southeastern end of the Temblor Range are similar to the shales of the Caliente Range west of the fault, indicating a 65-mile movement in 20 million years.

6 Staggered relationships of gravel beds in the San Emigdio and Temblor mountains show displacement of 10 miles in less than one million years.

7 The Garlock Fault has been offset about six miles from its western extension, the Big Pine Fault, during the past 200,000 years.

B. Later Miocene

A. Pre-Cretaceous

These two sketches show progressive land relationships, based on the San Andreas Fault movements shown above.

The Mechanics of an Earthquake

AN EARTHQUAKE MAY BE COMPLICATED, BUT IT IS NOT MYSTERIOUS. AND THE FACTS THAT LIE BEHIND THE RUMORS AND WILD STORIES ARE REALLY QUITE EASY TO UNDERSTAND.

WHEN THE EARTH MOVES SUDDENLY along a fault line, shock waves spread out in all directions, just as waves are created when you drop a pebble into a pond or when you tap the side of a bowl of gelatin with a spoon. But these are greatly oversimplified examples. The nature of an earthquake is extremely complex and involves masses and energies almost beyond comprehension.

The very size and strength of an earthquake is difficult to express in common terms. Some people have compared the force expended by a great earthquake to the force of 100,000 atomic bombs. But this comparison has very little real meaning since the force of even one such explosive is difficult to comprehend. Since the 1906 quake was the largest in California for which we have extensive records, its energy is often estimated. Two frequently quoted comparisons state that the energy released during that single earthquake was equal to the force needed to raise a cubic mile of rock 6,000 feet, or to run a battleship at full speed for 45,000 years. Since both of these conditions are hypothetical, they provide no real clue. It seems enough to say that an earthquake is one of the most irresistible forces on earth, and that it is impossible to attach any comparable numerical value on a force that can cause "solid" earth to shift 20 feet.

The shock waves that cause earthquakes are identifiable but complex. There are three general types:

The **P** wave is a sound wave. It travels at about 3½ miles a second and is the first to reach the surface. It is a longitudinal wave that creates a "push-pull' effect on the rock particles as it passes.

The **S,** or shear wave, travels about 2 miles a second near the surface and causes the earth to move in right angles to the direction of the wave. This is best illustrated by the action created when you snap a rope like a whip. The waves move the length of the rope, but the actual motion is at right angles.

The **L,** or long wave, is a slow surface wave that is usually distinguishable only at great distances. These long-period waves cause swaying of tall buildings and slight wave motion in bodies of water at great distances from the earthquake center.

These different waves show up as different patterns on a seismograph, but to the average person they mean only that the shocks created by a fault movement arrive at the surface at different times, so the result may be a period of motion that is greater than the actual duration of the fault movement. Many of the reports about "two separate shocks, a second apart" are really observations of the different arrival times of **P** and **S** waves. Twenty miles from the center of the earthquake the time lag will be about 4 seconds, and at a distance of 50 miles there will be an even more significant 10-second lapse.

Added to the complexities of earthquake motion are the unknown nature of the fault movement and the paths of the shock waves. There may be two or more fault movements, in quick succession, each with its own **P** and **S** waves that quickly become intermixed. The waves immediately begin to be reflected as they encounter obstacles in the earth, and the motion felt some distance from the source may consist not only of primary waves, but also of reflected and refracted shocks that may reinforce or cancel one another.

FORESHOCKS

Fault movements are seldom isolated events. In addition to the major shift, there frequently are minor movements on the same or related faults that result in foreshocks and aftershocks.

Foreshocks are difficult to recognize as such. They are caused by minor movements that sometimes precede and may even provide part of the triggering device for a main shock. They can occur weeks or months in advance, and their hypocenters may be somewhat removed from that of the main movement. Since there is no way of knowing with certainty the details of foreshocks, attempted analyses of minor earthquakes may be more confusing than revealing. No one can really tell whether a minor shock is just that or is the foreshock of a greater break to come. Even a series of light tremors that could be construed as the build-up of something greater may stop without ever being followed by a major earthquake.

It is only *after* an earthquake has run its course that seismologists can go back over the records and identify foreshocks. Unfortunately, no recognizable patterns have ever emerged. Several minor earthquakes from 1927 through 1932 can now be recognized as foreshocks to the disastrous Long Beach earthquake of 1933, but there were no known foreshocks of any kind preceding the 1906 and 1971 earthquakes.

AFTERSHOCKS

Aftershocks are much more common than foreshocks and are more predictable. Almost every major earthquake is followed by a series of aftershocks of varying strengths. Seismic records show that the number of aftershocks definitely increases with the strength of the main earthquake. They may continue for weeks, months or years, but the greatest concentration usually follows the main quake to cause panic among people who fear a repeat of the big one.

Aftershocks are a definite part of the release of accumulated elastic strain. When a great movement occurs, it not only relieves the pressure along its particular section of fault but it also changes the stress patterns for miles around. These changes, in turn, create adjustments of their own, and a chain effect results. When the earth is substantially distorted by an increase in pressure, it cannot be expected to return to its former condition in one movement. A simple comparison may be made by crumpling a sheet of stiff cellophane into a tight ball and holding it under pressure. When you release the pressure, the material springs back—but not very far. During the next few minutes, however, the material will continue to try to resume its former shape, with accompanying snaps and crackles, which may be compared to aftershocks.

Unfortunately, one or more of the aftershocks that follows a major earthquake may shake some towns hard enough to cause as much damage as the main earthquake.

SURFACE RUPTURES

Rarely is a fault movement strong enough to rupture the surface. Both the 1857 and the 1906 earthquakes along the San Andreas Fault were accompanied by extensive surface breaks, and there was both horizontal and vertical movement associated with the 1872 Owens Valley earthquake. Surface ruptures were associated with the quakes of 1940 in Imperial Valley, 1952 in Kern County, 1966 in Parkfield, 1966 in Imperial Valley, 1968 at Borrego Mountain, and 1971 at San Fernando. The 1966 earthquake in the Imperial Valley had a magnitude of only 3.6, and is the smallest known quake yet associated with surface displacement.

Primary surface displacements occur only along the fault line where the earthquake originated and should not be confused with cracks in highways and loose fill that show up miles away from the fault after a relatively strong earthquake. These are usually the result of earth slumps and landsliding rather than the tearing action of a fault movement.

DIAGRAM OF DISASTER—HOW AN EARTHQUAKE WORKS

This diagram illustrates the nature of earthquake motion and the type of surface damage that can result from the shock waves. The vertical scale has been distorted to provide a clearer picture.

The movement of rocks within the earth is the start-ing point of an earthquake. As the rocks snap past each other, great shock waves are generated that spread out in all directions; the strength of the waves is naturally higher near the point of movement. The earthquake that causes effects such as those shown here is a big one—

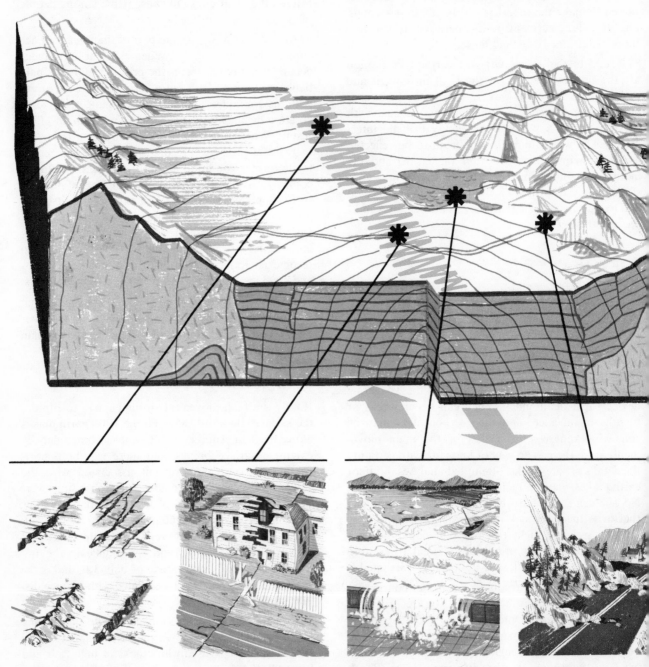

Evidence of surface movement includes long narrow ruptures, a series of parallel breaks, or pressure ridges that are "torn" in the earth. Low, steep scarps often are formed by vertical shifts.

Primary fault movement will rupture everything that crosses its path. Offsets in roads and fences often provide valuable clues in measuring the total amount of horizontal displacement.

Water in stream beds can actually be thrown completely over the banks. Shocks passing through lakes can create sizeable wave fronts that splash beyond regular shore lines.

Strong shock waves can uproot trees, snap off the of others, and cause mass landslides. However, stu houses on solid rock foundations may be able survive with minor dama probably magnitude 7.0 or greater.

There are two general types of surface disturbances during an earthquake. The first is the actual rupturing of the ground that is an expression of the basic shift in the rocks. This can take place only along the fault. The second is the wide-spread changes in the surface that are caused by the shock waves. The area affected by the spreading shock is impossible to define. But it is not uncommon for cities 20 miles from the fault to be hard hit, and for minor damage to occur in large buildings 80 to 100 miles away. The initial size of the earth movement, the geologic structure of the earth and quality of building construction are important factors in earthquake intensities, particularly at greater distances.

The shock waves shown here all emanate from a single point, where the first fault movement takes place. In reality, other movements may also occur along the same or related faults, each creating its own shock pattern and adding to the overall concentration and duration of the surface shaking.

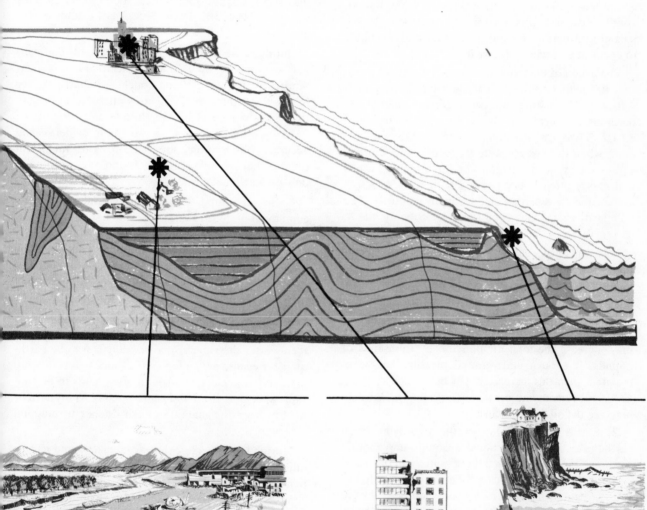

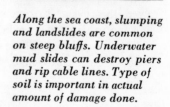

earthquake intensities tend to be highest in alluvium and soft soils. A town located in a river basin or flat alluvial plain may be badly damaged, while a similar city closer to the fault but constructed on bedrock mountains escapes unscathed. Large secondary slumps and lurches are common in farm areas, particularly where soil is water-laden, and are greater than primary movements on top of the fault.

At great distances, damage is usually limited to taller buildings that are swayed by slow surface waves. Damage may also occur when two buildings set up different motion and pound each other.

Along the sea coast, slumping and landslides are common on steep bluffs. Underwater mud slides can destroy piers and rip cable lines. Type of soil is important in actual amount of damage done.

RELATED PHENOMENA

Every major earthquake is followed by assorted stories about strange happenings. Usually these sidelights are based on a fact or a single true occurrence, but just as often, facts have been twisted and distorted into an imaginative tale. Most of the commonest phenomena can be reduced to something less than mysterious.

1. Earthquake Weather. Many residents of earthquake country are convinced that there is an "earthquake season," which is characterized by hot, humid, windless days. This is not true. Earthquakes are not partial to a particular kind of weather and are likely to occur at any time of the year.

The origin of the term "earthquake weather" probably lies with Aristotle, who taught that earthquakes were caused by winds that became imprisoned in underground caverns and shook the earth in trying to escape. Therefore, earthquakes occurred on windless days, when the breezes were trapped in the caverns. This association of earthquakes with escaping winds and gasses proved very popular with early scientists and the idea was perpetuated for many centuries. The popularity of the idea even today is probably due to its repetition in ancient literature.

One aspect of weather may have an indirect bearing on earthquakes. There is some evidence that the number of earthquakes in California generally increases in the fall months when large air masses associated with rain are on the move. The shifting of air fronts and the lowering of barometric pressure may increase the total air pressure by thousands of pounds. If enough underground pressure has accumulated, then this increase might be enough to "trigger" a movement and create an earthquake. But data here are definitely inconclusive.

2. Earth Fissures. Due to the treatment given earthquakes in some fictional accounts and motion pictures, it is often believed that fissures can open and swallow houses or even cities. This is completely untrue. Great earthquakes may open shallow cracks in the ground, and there may even be an opening and closing of the earth, but this is all on a very small scale. No one in California has ever dropped from sight into an earthquake fissure. During the 1906 quake, a cow was reportedly killed when it fell into a crack that momentarily opened in Marin County, but even this event is somewhat clouded by conflicting reports, and it may be that the animal actually fell into one of the trenches that are commonly formed when a great earthquake ruptures the surface.

3. Lights. In the records of great earthquakes, there are hundreds of reports of strange lights that are seen as the earth begins to shake. At one time, these lights, frequently described as "flashing" or "streaks of lightning," were thought to be related to the earthquake. But the reports became so persistent, even during earthquakes in clear daylight hours, that other reasons were sought.

Where earthquakes occur in heavily populated areas, seismologists have been able to trace many of the "strange lights" to electrical shorts and arcs. This was particularly true in the 1952 Kern County earthquakes.

But there are many other reports of lights that cannot be explained by weather or electricity. Investigations have proved nothing conclusive, but a theory has evolved that relates lights with static electricity that may be caused by the shifting and scraping of large rock formations. However, more evidence and accurate observations are needed. The important point is that the lights are effects and have nothing to do with earthquake causes.

4. Sounds. Reports of earthquake sounds are very common, but unfortunately, they do not always agree. There is a definite earthquake sound, however, that is quite separate from the rattling of dishes and the cracking of walls. It is a low-pitched moan or roar that is best heard in areas where the normal traffic noises of city life are absent.

The cause of the sound is the transfer of energy waves from the earth to the air during the first moments after the sub-surface grinding starts the earthquake. Because only a few of the shock waves fall into audible frequencies, the sound often tends to be sporadic and short-lived and gains much of its mysterious quality from the inability of the listener to pinpoint its origin.

During large quakes, the earthquake sound can be quite loud. Still, reports of its character frequently depend on the conditions surrounding the listener.

1872, Owens Valley: "Some described the noise as resembling that made by a whole park of artillery, shot off in rapid succession, with the rattling of musketry between."

1933, Long Beach: "A low rumbling noise resembling a heavy truck passing on the street."

1957, Daly City: "I thought it was a jet plane."

5. Earth Waves. One of the more frightening phenomenon of an earthquake is the appearance of earth waves, very similar to ocean swells, that race

across the ground. These usually are seen in sections where there is considerable loose alluvium or fill. Most reports include a note of surprise that such large waves did not rupture the ground. There is good reason—the ground waves are usually not as large as they seem to be. Ground waves definitely do exist, but observations are complicated by panic and loss of equilibrium of the observer as well as by air pressure changes just above the ground. Waves of 6 inches to 1 foot are not uncommon during very large earthquakes, but reports of 4 or 6-foot crests must be interpreted with some degree of caution.

6. Tidal Waves. The Japanese term *tsunami* is used internationally to describe ocean waves that sometimes accompany earthquakes. In California, tsunamis have never been too important. There were reports of a wave during the 1812 earthquake that may have reached 50-foot heights along the Santa Barbara coast (see page 140), but that is the only such significant incident on record. However, the largest earthquakes along other sections of the circum-Pacific belt, particularly in Chile, Japan, and the Aleutian Islands, have caused great tsunamis, and it may be that some future California earthquake will be accompanied by a tidal wave of some sort. Unfortunately, tsunamis are often the subject of wild rumor during earthquakes, so reports need careful checking. In general, any tsunami caused by a California quake will strike the shore immediately after the shock waves, so there can be little truth to post-earthquake reports you may hear about giant walls of water rushing in from the sea hours after the shaking has stopped.

7. Volcanoes. Mt. Lassen is the only recognized active volcano in California, and many small earthquakes that originate near it undoubtedly are related to volcanic activity. But earthquakes that originate along fault lines have no connection with volcanoes. A few quakes in Owens Valley and the desert area near the Mexican border may be related to dying volcanic influence. But these are exceptions. The reports of "volcanic craters" that appear during a major earthquake are usually exaggerations of minor eruptions of sand and water that form minor craterlets. These sand and water spouts form surface deposits up to 5 or 6 feet in diameter. Their action ends as soon as the shaking stops and the underground water settles down. Where a water spout continues to flow for some time after an earthquake, a considerable change in the water table has taken place and created a change in pressure and forced water to the surface. The flow normally subsides within a few hours.

8. Actions of Animals. If an earthquake is troubling and mysterious to humans, it is doubly so to animals. Dogs, cats, and horses frequently can feel minor tremblings and small foreshocks that escape human senses, and they become uneasy for no apparent reason. During an actual earthquake, animals often panic and react blindly. When the shaking stops, they usually are fearful of the spots where they were standing when the earthquake struck.

Such simple and normal reactions lead to strange stories about the mysterious sixth sense of animals that warns them of impending doom. In reality, the reactions are quite in harmony with the animals' sharp senses, general fear of the unknown, and tendency to act according to instinct.

Great flocks of birds have been known to take off from trees and bushes just before an earthquake. Once again, a special sixth sense is supposed to have warned them. But there is another explanation. Earthquakes can do great damage to trees—tearing down limbs and snapping off trunks—and even the slightest tremor will cause a general shaking of the branches. It is this trembling, often caused by small foreshocks, felt in trees and bushes an instant before the ground actually begins to shake, that startles birds into flight.

Fish, however, have nowhere to go, and the damage can be disastrous. Earthquake shock waves differ little from those caused by an underwater explosion, and a major quake can kill hundreds of fish.

The 1906 earthquake caused step-down dislocations in soft ground near Salinas. This type of secondary effect in loose fill gives a distorted picture of earthquake force.

How Earthquakes Are Measured

AN EARTHQUAKE CAN BE MEASURED BY INSTRU-
MENTS OR ACTUAL OBSERVATIONS. BUT OFTEN,
THE RAW STATISTICS BEAR LITTLE RELATION
TO THE PANIC AND CONFUSION THAT ARE CRE-
ATED WHEN THE EARTH ROCKS.

EARTHQUAKES CAN BE MEASURED in terms of either energy (magnitude) or actual effects (intensity). The first measurement is based on instrument records, the second on personal observations. They are completely separate in intent and results, but they are often confused by the general public.

THE RICHTER MAGNITUDE SCALE

The Richter Magnitude Scale measures the size of an earthquake at its source. The measurements are based on records made on a standard type of seismograph at a distance of 62 miles (100 km) from the epicenter. Seismograms from several different stations are normally used in computing the magnitude of a quake. Since most stations are bound to be some distance from the source other than the standard 62 miles, many records are compared and complex conversion tables are used to arrive at the final figure.

However, seismologists can make a fairly accurate estimate of magnitude within minutes after the earthquake by carefully examining the record taken at only one seismographic station. It is this figure that newspapers report in their first stories about a quake. In recent years, magnitudes have become an indispensable part of newspaper reports about earthquakes, and readers have learned to watch for them as the only expressions of earthquake strength that are understandable. Magnitudes are expressed in whole numbers and decimals—usually between 3 and 8—so they are easy to grasp and remember.

One of the most popular uses made of the Richter magnitudes is in the comparison of earthquake size. Every newspaper story invariably includes some statement to the effect that ". . . the magnitude of yesterday's earthquake was 4.3. The magnitude of the 1906 earthquake was 8.3." However, these figures can be very misleading unless you understand the mathematical basis for the Richter scale. There are two important points to remember. First, the maximum amplitude of earthquake waves recorded on the seismogram is transformed to a numerical figure by means of a logarithmic scale. This means that an increase of a whole number on the scale represents a *tenfold* increase in the size of the earthquake record. So the record written by an earthquake of 8.3 magni-

DELICATE INSTRUMENTS RECORD THE EARTH'S VIOLENCE

Any attempt to record earthquake movement is handicapped by the necessity of resting the measuring instrument (seismograph) on the very earth that is moving. Because the base must be anchored in rock, there is no way of setting up an instrument that can operate completely independent of earth movement. However, there are ways of establishing a steady mass that will remain *relatively* inert when the earth moves beneath it, thereby creating a significant, measurable difference between its motion and that of the earth. This mass is usually a pendulum. We are accustomed to watching a pendulum move while the earth is at rest; but in a seismograph, the opposite is true. The pendulum is suspended either vertically or horizontally as shown below. The weight is joined to a supporting pillar by only a wire and a boom resting in a universal joint, so that the delicately-suspended mass will tend to hold its original position even though the support vibrates considerably during an earthquake. Thus, even large earth movements result in only minute shifts of the pendulum.

These small movements must then be converted into a visible record. The recording device may be either a very sensitive pen that is attached directly to the pendulum, or a beam of light that is reflected off a mirror, as shown below. The recording medium is blank paper or, in the case of light beam recording, photographic paper that is attached to a rotating drum. Time segments are marked on the paper, so that the arrival time of the earthquake waves can be accurately pinpointed.

A seismograph of this type can record only one type of motion, at right angles to the length of the pendulum. So to obtain a truly accurate record, each recording station must have three seismographs—one each for north-south, east-west, and vertical motion.

Recent improvements in seismographs have made them more valuable in recording very weak motion, and in determining the exact nature and direction of the various wave forms. Most modern seismographs use an electrical pickup of pendulum movements, and transfer this current through a magnifying galvanometer. With this arrangement, slight motion can be magnified many times to make a more legible record.

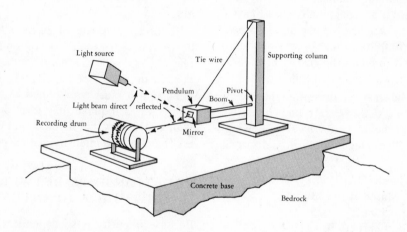

The record written on a seismograph is called a seismogram. It is a continuous line that fluctuates according to the pendulum movements. The greater the earth movement, the greater the variation in the record. The degree of amplitude is the basis for assigning a Richter magnitude to the shock. The arrival time of the P, S and L waves enables seismologists to fix the distance between epicenter and the recording seismograph. By comparing the records taken at several different stations, the source of the waves can be accurately pinpointed in both direction and distance.

The seismogram shown below was taken at the University of California Seismographic Station in Berkeley. Source of the waves was an earthquake of magnitude 6½ that originated in the Aleutian Islands, 2,550 miles from Berkeley, shortly after noon on July 13, 1959.

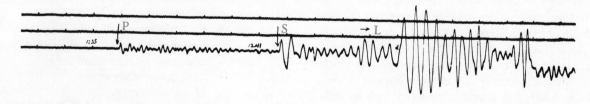

tude is not twice that of a shock of 4.3 magnitude, but 10,000 times as great. Does this great difference mean that many types of seismographs are required to record all the earthquakes of various sizes that occur in California? Yes it does, and since there is a great difference in sensitivity between the strong-motion and weak-motion instruments, any magnitude comparisons must be made judiciously.

Second, the energy released at the source of earthquakes of different magnitudes is even more variable than the seismographic records they create. Again owing to the structure of the scale, an increase of one whole number indicates an energy release about 60 times greater than that of the next lower number. Therefore, a magnitude 8.3 earthquake generates about 10,000,000 times as much energy as a magnitude 4.3 shock.

That these facts about the Richter Magnitude Scale are not clearly understood was emphasized after the 1957 Daly City earthquake. Some publications seriously compared the building damage caused by the earthquake, which had a 5.3 magnitude, with that of the 1906 earthquake, which had 8.3 magnitude. Since the energy of at least 50,000 earthquakes of magnitude 5.3 would be required to equal the energy of an 8.3 shock, the comparison was somewhat ludicrous.

The Richter Scale has no fixed maximum; however, observations have placed the largest known earthquakes in the world at the 8.8 or 8.9 level. Following are some representative magnitudes of California earthquakes.

5.3 Daly City, March 22, 1957
6.3 Long Beach, March 10, 1933
6.3 Santa Barbara, June 29, 1925
6.6 San Fernando, February 9, 1971
7.7 Kern County, July 21, 1952
8.3 Northern California, April 18, 1906

The Richter Magnitude Scale was first published in 1935. Past records have made it possible to assign magnitudes to major earthquakes as far back as 1904. In the following chapters, magnitudes are given whenever available.

When analyzing magnitudes, keep in mind that they provide only an index of potential energy. The instrumental computation does not take into effect location or depth of hypocenter, or ground and structural conditions in the affected area. Therefore, Richter numbers cannot be used to estimate damage. An earthquake that is centered in a densely populated area, and which kills dozens and levels a city, may have exactly the same magnitude as an earthquake far out in the desert that does nothing more than raise a cloud of dust and frighten a few rabbits.

THE MODIFIED MERCALLI INTENSITY SCALE

This scale, which is printed on another page, is much more meaningful to laymen, since it is based on actual observations of earthquake effects at specific points. But since that data used for assigning intensities can be gathered only from firsthand reports, weeks or months are sometimes required before an intensity report can be issued.

While an earthquake can have only one magnitude, it can have several intensities. The intensity is highest near the epicenter, and it gradually decreases as distance from the epicenter increases. For example, the 1957 Daly City earthquake had an intensity rating of VII near the epicenter near Mussel Rock, VI in some parts of San Francisco, IV in other sections of the same city, and something between I and II at Stockton.

If the comparisons of earthquake intensity are to be accurate, then each observation must be a personal one. If you decide to send in an intensity report on the next earthquake (see page 159), then make up your mind that it must be confined to your own observations. An earthquake may level dozens of houses in your community, but if you do not feel even the slightest shaking, the intensity at that particular spot is I.

RELATIONSHIPS BETWEEN MAGNITUDE AND INTENSITY

Because intensity is so dependent on the particular ground and structural conditions of a particular area, it may vary considerably at two points that are equidistant from an epicenter. For this reason, it is difficult to equate magnitude with estimated intensity. However, the following brief summary provides a general estimate of the relationships:

MAGNITUDE	EFFECTS (INTENSITY)
1	Only observed instrumentally
2	Can be barely felt (Intensity II) near epicenter
4.5	Felt to distances of some 20 miles from the epicenter; may cause slight damage (Intensity VII) in small area
6+	Moderately destructive
7+	Major earthquake
8+	Great earthquake (1906, 1857, 1872)

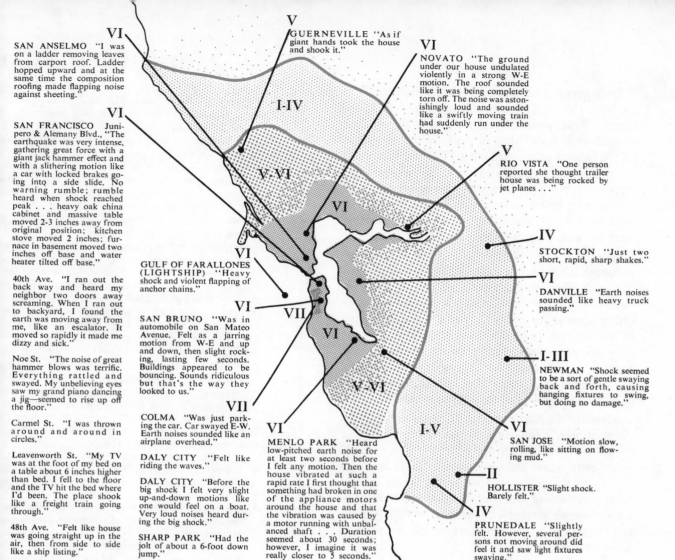

V
GUERNEVILLE "As if giant hands took the house and shook it."

VI
SAN ANSELMO "I was on a ladder removing leaves from carport roof. Ladder hopped upward and at the same time the composition roofing made flapping noise against sheeting."

VI
NOVATO "The ground under our house undulated violently in a strong W-E motion. The roof sounded like it was being completely torn off. The noise was astonishingly loud and sounded like a swiftly moving train had suddenly run under the house."

VI
SAN FRANCISCO Junipero & Alemany Blvd., "The earthquake was very intense, gathering great force with a giant jack hammer effect and with a slithering motion like a car with locked brakes going into a side slide. No warning rumble; rumble heard when shock reached peak . . . heavy oak china cabinet and massive table moved 2-3 inches away from original position; kitchen stove moved 2 inches; furnace in basement moved two inches off base and water heater tilted off base."

V
RIO VISTA "One person reported she thought trailer house was being rocked by jet planes . . ."

40th Ave. "I ran out the back way and heard my neighbor two doors away screaming. When I ran out to backyard, I found the earth was moving away from me, like an escalator. It moved so rapidly it made me dizzy and sick."

IV
STOCKTON "Just two short, rapid, sharp shakes."

VI
GULF OF FARALLONES (LIGHTSHIP) "Heavy shock and violent flapping of anchor chains."

VI
DANVILLE "Earth noises sounded like heavy truck passing."

Noe St. "The noise of great hammer blows was terrific. Everything rattled and swayed. My unbelieving eyes saw my grand piano dancing a jig—seemed to rise up off the floor."

VII

VI
SAN BRUNO "Was in automobile on San Mateo Avenue. Felt as a jarring motion from W-E and up and down, then slight rocking, lasting few seconds. Buildings appeared to be bouncing. Sounds ridiculous but that's the way they looked to us."

I-III
NEWMAN "Shock seemed to be a sort of gentle swaying back and forth, causing hanging fixtures to swing, but doing no damage."

Carmel St. "I was thrown around and around in circles."

VII
COLMA "Was just parking the car. Car swayed E-W. Earth noises sounded like an airplane overhead."

Leavenworth St. "My TV was at the foot of my bed on a table about 6 inches higher than bed. I fell to the floor and the TV hit the bed where I'd been. The place shook like a freight train going through."

DALY CITY "Felt like riding the waves."

DALY CITY "Before the big shock I felt very slight up-and-down motions like one would feel on a boat. Very loud noises heard during the big shock."

VI
MENLO PARK "Heard low-pitched earth noise for at least two seconds before I felt any motion. Then the house vibrated at such a rapid rate I first thought that something had broken in one of the appliance motors around the house and that the vibration was caused by a motor running with unbalanced shaft . . . Duration seemed about 30 seconds; however, I imagine it was really closer to 5 seconds."

VI
SAN JOSE "Motion slow, rolling, like sitting on flowing mud."

48th Ave. "Felt like house was going straight up in the air, then from side to side like a ship listing."

SHARP PARK "Had the jolt of about a 6-foot down jump."

II
HOLLISTER "Slight shock. Barely felt."

IV
PRUNEDALE "Slightly felt. However, several persons not moving around did feel it and saw light fixtures swaying."

Reports on the 1957 Daly City earthquake gathered by the U.S. Coast and Geodetic Survey indicate the variety of personal sensations that can result from even a minor tremor.

MEASURING EARTHQUAKES— THE HUMAN FACTOR

Measured seismologically, the Daly City earthquake on March 22, 1957 was not a large one. The Richter magnitude was only 5.3 and the intensity along the coastline near Mussel Rock barely reached VII. But when an earthquake strikes a densely populated area such as the Bay Area, the impact of the shock frequently goes beyond the cold statistics.

When the floor of your living room sways sickeningly like a ship at sea and the heavy furniture jiggles crazily, it is often difficult to find reassurance in the fact that the force behind the movement is only one-thousandth that of the 1906 earthquake.

Earthquake damage to buildings is fairly predictable; not so with humans. Even so astute an observer as Charles Darwin was impressed by the effect that an earthquake can have on the senses. In 1835, after experiencing a large shock in Chile, Darwin made this entry in his diary; "A bad earthquake at once destroys our oldest associations; the earth, the very emblem of solidity, has moved beneath our feet like a thin crust over a fluid;—one second of time has created in the mind a strange idea of insecurity, which hours of reflection would not have produced."

This feeling apparently is universal, even when property damage is comparatively slight, as in 1957. Personal accounts of this and other earthquakes of equal size show that human nerves sometimes give way faster than the weakest of buildings. An earthquake that causes only a few plaster cracks and broken store fronts can send hysterical women screaming into the streets and freeze strong men in their office chairs in momentary fear.

Since personal reactions and observations of damage form the bases for assignments of earthquake intensity, these personal feelings can often obscure analysis of a shock. Most of the information offered by the general public, including all of the statements on the illustration on page 33, are reasonable personal accounts. But it is not uncommon for seismologists to receive long letters describing gaping earth fissures, huge earth waves, and wildly gyrating buildings that seem to touch the ground with each giant swing.

After long examination of reports of personal experiences, it has become quite evident to seismologists that both physical and mental states can affect an observation. For example, people lying on the floor or sitting down at home may feel definite shaking during an earthquake, while others walking around in the same room may not even be aware of the slightest tremors. And the man who is sitting in an automobile, being rocked by its suspension system, is very likely to get a distorted view of the world around him.

Mentally, the distortions may be even greater. When the earth moves beneath your feet, it can upset your normal body equilibrium and affect all of your senses. Small earth waves suddenly loom as large as waves rolling up the beach, and the world seems to turn topsy-turvy before your eyes. The sounds of an earthquake become mingled with the crash of falling dishes and the roar of traffic until the total volume of noise seems overwhelming.

Compounding this temporary mental confusion is the almost universal feeling of fear that a great earthquake can cause. Investigations into this phenomenon indicate that the fear does not arise just from the potential damage that can be done. But rather, it is more concerned with the unknown qualities of an earthquake. Here is an invisible force, rising out of the ground, that is of unpredictable strength and size. Even after the shaking starts, there is no way of knowing whether this is the biggest shock of a series, or whether it is a relatively small prologue to a real "monster."

And to watch buildings crack and fall, or trees shake, without being able to see the cause behind the effect is a terrifying sensation.

The panic that accompanies a great earthquake is one of the greatest hazards to rescue work and the rapid assessment of damage. However, there seems to be no pat, reassuring phrases that will comfort a woman who has been knocked flat on her back and pinned down by an earthquake that has suddenly turned her backyard into a roller coaster ride.

The first scale developed to indicate the varying inter of earthquake shock was developed in Europe in the by De Rossi of Italy and Forel in Switzerland. The Forel Scale, with values from I to X, was widely use about two decades as a means of investigating earthq and comparing the effects of various shocks throu the world. The scale's main defect was that it lump great deal of major damage under classification X. Thi fine during the early stages of technology, but a science of seismology progressed, the need for a refined scale was greatly increased.

In 1902, the Italian seismologist Mercalli set up a scale, which was based on a I to XII range and pro for more refined analysis of major damage. The M Scale was modified in 1931 by two American seismolo Harry O. Wood and Frank Neumann, to take into ac modern features such as tall buildings, motor cars trucks, and underground water pipes. It is this Mo Mercalli Scale (frequently abbreviated to MM) that used today. In the version printed here, the languag been slightly changed, but the basic ideas are the as those used by professional seismologists to rate quakes.

The varying intensity grades of an earthquake frequ are expressed in an isoseismal map, with roughly ci lines drawn through areas of equal intensity. Several maps appear in the following chapters.

All intensity figures used in this book are MM Rossi-Forel numbers given to earthquakes before 1900 been adjusted to fit the more modern Mercalli syste

I **Not felt by people, except under especially favo circumstances. However, dizziness or nausea ma experienced.**
Sometimes birds and animals are uneasy or distu Trees, structures, liquids, bodies of water may gently, and doors may swing very slowly.

II **Felt indoors by a few people, especially on upper of multistory buildings, and by sensitive or ne persons.**
As in Grade I, birds and animals are disturbed trees, structures, liquids and bodies of water may Hanging objects swing, especially if they are delic suspended.

III **Felt indoors by several people, usually as a rapid tion that may not be recognized as an earthqua first. Vibration is similar to that due to passing of a or lightly loaded trucks, or heavy trucks some dis away. Duration may be estimated in some cases.**
Movements may be appreciable on upper levels structures. Standing motor cars may rock slightly.

IV **Felt indoors by many, outdoors by few. Awakens individuals, particularly light sleepers, but frighter one except those apprehensive from previous ex ence. Vibration like that due to passing of heav heavily loaded trucks. Sensation like a heavy striking building, or the falling of heavy objects in** Dishes, windows and doors rattle; glassware and c ery clink and clash. Walls and house frame c especially if intensity is in the upper range of this g

Hanging objects often swing. Liquids in open vessels are disturbed slightly. Stationary automobiles rock noticeably.

V Felt indoors by practically everyone, outdoors by most people. Direction can often be estimated by those outdoors. Awakens many, or most sleepers. Frightens a few people, with slight excitement; some persons run outdoors.

Buildings tremble throughout. Dishes and glassware break to some extent. Windows crack in some cases, but not generally. Vases and small or unstable objects overturn in many instances, and a few fall. Hanging objects and doors swing generally or considerably. Pictures knock against walls, or swing out of place. Doors and shutters open or close abruptly. Pendulum clocks stop, or run fast or slow. Small objects move, and furnishings may shift to a slight extent. Small amounts of liquids spill from well-filled open containers. Trees and bushes shake slightly.

VI Felt by everyone, indoors and outdoors. Awakens all sleepers. Frightens many people; general excitement, and some persons run outdoors.

Persons move unsteadily. Trees and bushes shake slightly to moderately. Liquids are set in strong motion. Small bells in churches and schools ring. Poorly built buildings may be damaged. Plaster falls in small amounts. Other plaster cracks somewhat. Many dishes and glasses, and a few windows, break. Knick-knacks, books and pictures fall. Furniture overturns in many instances. Heavy furnishings move.

VII Frightens everyone. General alarm, and everyone runs outdoors.

People find it difficult to stand. Persons driving cars notice shaking. Trees and bushes shake moderately to strongly. Waves form on ponds, lakes and streams. Water is muddied. Gravel or sand stream banks cave in. Large church bells ring. Suspended objects quiver. Damage is negligible in buildings of good design and construction; slight to moderate in well-built ordinary buildings; considerable in poorly built or badly designed buildings, adobe houses, old walls (especially where laid up without mortar), spires, etc. Plaster and some stucco fall. Many windows and some furniture break. Loosened brickwork and tiles shake down. Weak chimneys break at the roofline. Cornices fall from towers and high buildings. Bricks and stones are dislodged. Heavy furniture overturns. Concrete irrigation ditches are considerably damaged.

VIII General fright, and alarm approaches panic.

Persons driving cars are disturbed. Trees shake strongly, and branches and trunks break off (especially palm trees). Sand and mud erupts in small amounts. Flow of springs and wells is temporarily and sometimes permanently changed. Dry wells renew flow. Temperature of spring and well waters varies. Damage slight in brick structures built especially to withstand earthquakes; considerable in ordinary substantial buildings, with some partial collapse; heavy in some wooden houses, with some tumbling down. Panel walls break away in frame structures. Decayed pilings break off. Walls fall. Solid stone walls crack and break seriously. Wet ground and steep slopes crack to some extent. Chimneys, columns, monuments and factory stacks and towers twist and fall. Very heavy furniture moves conspicuously or overturns.

IX Panic is general.

Ground cracks conspicuously. Damage is considerable in masonry structures built especially to withstand earthquakes; great in other masonry buildings—some collapse in large part. Some wood frame houses built especially to withstand earthquakes are thrown out of plumb, others are shifted wholly off foundations. Reservoirs are seriously damaged, and underground pipes sometimes break.

X Panic is general.

Ground, especially when loose and wet, cracks up to widths of several inches; fissures up to a yard in width run parallel to canal and stream banks. Landsliding is considerable from river banks and steep coasts. Sand and mud shifts horizontally on beaches and flat land. Water level changes in wells. Water is thrown on banks of canals, lakes, rivers, etc. Dams, dikes, embankments are seriously damaged. Well-built wooden structures and bridges are severely damaged, and some collapse. Dangerous cracks develop in excellent brick walls. Most masonry and frame structures, and their foundations, are destroyed. Railroad rails bend slightly. Pipe lines buried in earth tear apart or are crushed endwise. Open cracks and broad wavy folds open in cement pavements and asphalt road surfaces.

XI Panic is general.

Disturbances in ground are many and widespread, varying with the ground material. Broad fissures, earth slumps, and land slips develop in soft, wet ground. Water charged with sand and mud is ejected in large amounts. Sea waves of significant magnitude may develop. Damage is severe to wood frame structures, especially near shock centers; great to dams, dikes and embarkments, even at long distances. Few if any masonry structures remain standing. Supporting piers or pillars of large, well-built bridges are wrecked. Wooden bridges that "give" are less affected. Railroad rails bend greatly, and some thrust endwise. Pipe lines buried in earth are put completely out of service.

XII Panic is general.

Damage is total, and practically all works of construction are damaged greatly or destroyed. Disturbances in the ground are great and varied, and numerous shearing cracks develop. Landslides, rock falls, and slumps in river banks are numerous and extensive. Large rock masses are wrenched loose and torn off. Fault slips develop in firm rock, and horizontal and vertical offset displacements are notable. Water channels, both surface and underground, are disturbed and modified greatly. Lakes are dammed, new waterfalls are produced, rivers are deflected, etc. Surface waves are seen on ground surfaces. Lines of sight and level are distorted. Objects are thrown upward into the air.

How and Why Earthquakes Cause Damage

BY ITSELF, A QUAKE CAN DO LITTLE DAMAGE. BUT WHEN IT STRIKES AN AREA WHERE MAN HAS ERECTED FLIMSY BUILDINGS ON LOOSE SOIL, EVEN A MODERATE SHOCK CAN CAUSE INJURY, DEATH AND DESTRUCTION.

BUILDING DAMAGE presents some of the best and worst evidence for use in observing and measuring an earthquake. The fact that the damage is permanent makes it ideal for detailed examination. Although personal impressions of an earthquake may become distorted by the excitement of the moment and natural phenomena such as ground waves and sounds pass in an instant, the damage to buildings remains and can be studied and analyzed after the confusion is over.

But building damage usually tells much more about the structures than it does about the earthquake. The extent of the damage depends more on the strength of a building and the ground beneath it than it does on the strength of the shock waves. (An excellent reference work on this subject is *Earthquake Hazard in the San Francisco Bay Area: A Continuing Problem in Public Policy,* written by Karl V. Steinbrugge and published by the University of California Institute of Governmental Studies in 1968).

Buildings suffer during an earthquake primarily because horizontal forces are exerted against structural designs that often are meant to contend with only vertical stresses. A further complicating factor is the uneven resistance of different parts of a house—a slab floor can absorb much more shock force than an unbroken large pane of glass. When the give-and-take of any structural element is unbalanced, the rigid, weaker elements are fractured or torn loose. The classic example of this type of damage is the brick chimney that cannot move with the same flexibility as a wood frame house, and is therefore snapped off in an earthquake, usually at the roof line.

There are five principal elements that influence damage to man-made structures:

1. Strength of the earthquake waves reaching the surface. A relatively weak fault movement near the earth's surface, or a giant shift far underground, usually cannot generate the initial strength required to overcome the natural built-in strength of even the weakest structures. But an earthquake with magnitude 4.5 or greater, originating from a depth of some 10 miles (where most California earthquakes originate) usually results in maximum intensities of VI, VII, or VIII, which can include substantial building damage.

2. Length of the earthquake motion. Very rarely will an earthquake shock be felt as a single pulse; it

Damage at the Veterans Administration Hospital in Sylmar during the 1971 San Fernando earthquake presented a classic picture of inadequate vs. modern building techniques. The center structures, built in 1926, collapsed "like smashed orange crates," according to one observer. The outer structures, built in 1937 and 1947 after earthquake-resistant designs were included in building codes, escaped without any significant structural damage. The death toll at the hospital was 45, and would have been much higher if the quake had struck during normal business hours.

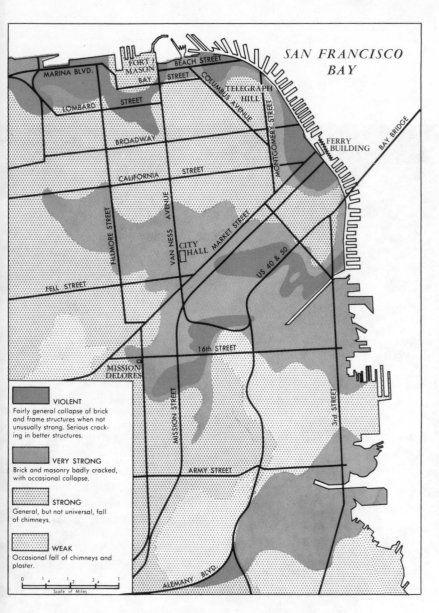

Map labels: SAN FRANCISCO BAY, FORT MASON, BAY, MARINA BLVD., BEACH STREET, STREET, COLUMBUS AVENUE, TELEGRAPH HILL, LOMBARD, STREET, BROADWAY, MONTGOMERY STREET, FERRY BUILDING, BAY BRIDGE, CALIFORNIA, STREET, FILLMORE STREET, VAN NESS AVENUE, CITY HALL, MARKET STREET, US 40 & 50, FELL STREET, 16th STREET, MISSION DELORES, 3rd STREET, MISSION STREET, ARMY STREET, ALEMANY BLVD.

VIOLENT
Fairly general collapse of brick and frame structures when not unusually strong. Serious cracking in better structures.

VERY STRONG
Brick and masonry badly cracked, with occasional collapse.

STRONG
General, but not universal, fall of chimneys.

WEAK
Occasional fall of chimneys and plaster.

0 ¼ ½ ¾ 1
Scale of Miles

The importance of solid rock foundations in minimizing earthquake damage was clearly illustrated by the pattern of destruction in San Francisco during the 1906 earthquake. Building damage was consistently greatest on land that had been recovered from the bay or filled in over old swamps and river beds. On the higher bedrock hills, intensities were considerably less. Proximity to the fault was much less important; for example, the area near the Ferry Building was farthest from the break, but was hardest hit because of unfavorable ground conditions. Uncompacted fill also lurched and settled unevenly, causing streets to crack and buildings to sway drunkenly off their foundations.

is usually a fluctuating series of tremors that last from 10 seconds to a minute. It is the cumulative effect of this motion that works on structural walls and is the usual cause of collapse.

Aftershocks play a very important role in the amount of building damage attributed to an earthquake. The main shock may substantially weaken many buildings without actually knocking them down, then a relatively weak aftershock may bring them down. This was vividly demonstrated in the Kern county earthquakes of 1952, when an aftershock of only 5.8 magnitude that came a full month after the July 21 main earthquake did most of the damage to Bakersfield. Since many buildings had been substantially weakened by the main quake, the aftershock struck a very vulnerable spot.

Many property owners do not fully repair their buildings after a major earthquake, thereby setting the stage for substantial and surprising damage the next time a weak earthquake strikes the same spot. "Paint and plaster" can cover the weakened joints and sagging frames, but the building may unexpectedly collapse years later when subjected to even minor strain.

3. Proximity to the fault. This is important only as a general concept. Obviously, a house 5 miles from the fault that causes an earthquake is in more danger than a house 100 miles away. But distance is not necessarily a direct relationship. Unless you live right on top of a fault zone, proximity (within 10 or 20 miles) is much less important than the construction of your home and the geologic foundation under it. During the 1906 earthquake, for example, the city of

Santa Rosa was much harder hit than surrounding towns, even though it is located some 20 miles from the San Andreas Fault. The reasons behind the heavy damage were traced to structural inadequacies and weak ground.

4. Geologic foundation. To many engineers and insurance experts, this is perhaps the most important factor in building damage. Earthquake studies have almost invariably shown that the intensity of a shock is directly related to the type of ground supporting the building. Structures built on solid rock near the epicenter of an earthquake frequently fare better than more distant buildings on soft ground.

Fill and "made" land, especially when water-soaked, is known to transmit much greater intensity of motion than rock outcroppings located nearby. Loose, water-charged natural ground is also dangerous. Even worse than building on soft ground is locating a structure partly on rock or hard ground and partly on soft ground. Differential settlement can impose tremendous strain on the structure that may prove disastrous during an earthquake.

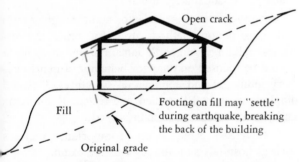

If fill used in hillside lots is not properly compacted, earthquake may cause it to settle, and the resulting stress can break the back of even a well-built house.

Modern developments in foundation engineering have partially reduced the hazard in some types of fill. During the 1957 Daly City earthquake, for example, it was found that engineered, recompacted soil (fill, sometimes 35 feet deep) under modern subdivisions was not a significant factor in damage.

5. Building design. Architects and engineers for years have declared that any building can be designed to be earthquake resistant, provided its site is suitable. There is no mystery to the requirements. The object is to insure that the structure will overcome inertia and move with the earth as a unit, not as an unrelated assembly of parts. The basic essential is strength, obtained by adequate bracing and structural continuity,

with secure anchoring and bonding of all elements—foundation, frame, outer and inner walls, upper floors, roof. In short, a building must be well balanced and tied together.

Conventional wood frame construction has a high degree of quake resistance, provided the workmanship is sound. The frame, roof and wall sheathing, and interior wall and ceiling panels supply the necessary bracing.

The modern post-and-beam house with walls of glass, however, is a different matter. Lacking the built-in bracing of the conventional frame house, the post-and-beam design with its potential "hinge" connections of uprights and crosspieces must be carefully designed to resist horizontal force. Even then, the large glass areas remain a problem. Large windows in homes and stores are often shattered in earthquakes, unable in their tightly fixed positions to withstand the twisting they receive. Engineers are studying ways to mount such panes of glass with some freedom of movement to enable them to "give."

Ordinary masonry structures have been especially vulnerable to earthquake damage. This a result of their heavy mass, as compared with that of wood frame houses. However, where good brick or concrete block has been laid with strong mortar, adequately reinforced with steel, and properly tied to foundations and inner walls, masonry buildings have resisted quakes very well. Concrete slab floors offer no special quake problems when the concrete is of good quality, the soil beneath the slab is sound, and the footings are secure.

If your community has a building code that conforms with the earthquake provisions of the uniform building code now widely in use—and if your building inspector has the experienced staff to enforce the code —then you can be sure your design, when approved, is relatively safe. This is no guarantee that a house conforming to the code won't suffer damage in an earthquake. It means that such a house can be considered safe from the standpoint of injury to occupants, and that it is designed to minimize danger. Keep in mind, too, that building codes do nothing more than set minimum standards, so just meeting the requirements may mean that your house is on the verge of being unsafe. In many cases, an architect or structural engineer can increase the earthquake resistance without increasing the cost substantially.

Choice and quality of materials are very important, and specifications should be rigidly met. Weak mortar and poor lumber are no match for the assault of an earthquake. Of equal importance is the workmanship required in transforming the materials into a complete structure.

Tall buildings have certain characteristics which make them particularly susceptible to damage from the slow and gentle rocking motion of a major distant earthquake as much as 100 to 200 miles away. Recent earthquakes (particularly 1952 in Kern County) have amply demonstrated that this gentle oscillation of the earth's surface can cause quasi-resonance to buildings which are five or more stories high.

SLOW ACCEPTANCE OF BUILDING CODES

Scientific understanding of earthquake damage and its causes has improved considerably during the last half century. However, legislation and adoption of building codes have not kept pace. Despite the thorough study and excellent documentation of the 1906 earthquake, major quakes coming decades later (notably Santa Barbara in 1925 and Long Beach in 1933) revealed some shocking building practices that indicated either lack of knowledge or conscience concerning the effect of earthquakes on structures. The 1933 quake focused special attention on public schools. With few exceptions, school buildings over a wide area lost walls or roof, or simply collapsed.

The result of this dramatic chain of disasters was a sudden public concern that prompted the California legislature to pass the "Field Act," which gave the State Division of Architecture authority and responsibility for approving design and supervising construction of public schools and establishing severe penalties for violations. The 1933 earthquake also pointed up the inadequate provisions and poor enforcement of many local building codes, especially as they applied to large public buildings. Quakes of recent years have provided many graphic contrasts between buildings that meet modern code requirements and older structures built with no provision for earthquake resistance.

SAN FERNANDO—THE REAL TEST

The 1971 San Fernando earthquake provided a real test of supposedly earthquake-resistant construction in California. In general, modern buildings fared much better than older buildings in areas of moderately strong ground shaking. But many structures failed the test, particularly in the areas of very strong ground motion. Hospitals were hard hit, freeway bridges collapsed, and residential damage was heavy.

Overall, more than 700 buildings were evacuated and declared unsafe, and an additional 8,000 received varying degrees of damage.

Of the school buildings located in the stricken area, those built after passage of the Field Act suffered virtually no damage at all, except for nonstructural failures (damaged ceilings and light fixtures) where ground vibrations were particularly strong.

In the preliminary report on the quake published by the U.S. Geological Survey and the National Oceanic and Atmospheric Administration, the Earthquake Engineering Research Institute Committee expressed these conclusions regarding construction:

1. The hazard of non-earthquake-resistive construction was reemphasized by the quake. Old structures, particularly unreinforced masonry bearing wall type with weak mortar, performed poorly. Collapse of skeleton concrete frame buildings with unreinforced hollow tile walls at the Veterans Hospital resulted in the greatest loss of life.

2. Modern high-rise construction survived well. However, most were too far from the quake's epicenter to be adequately tested.

3. A reassessment is needed on some design criteria and construction methods used for bridges, overpasses, buildings, and earth dams. Particular attention should be focused on the poor behavior of many one-story industrial and commercial structures. Plywood diaphragms must be studied and tested. Reinforcing and joinery details must be improved.

4. Increased seismic safety of important structures such as hospitals and utilities must be examined.

5. Greater caution is required relative to building in zones of active faults.

6. Modern one-story wood frame and stucco dwellings performed reasonably well considering the high ground accelerations and movements. However, reconsideration of code provisions and enforcement is indicated, particularly for two-story construction and masonry chimneys.

Other observers of the San Fernando quake thought these recommendations were far too conservative, and spoke out in favor of an intensive reevaluation of the state's basic design criteria and building techniques. It is quite obvious from this quake that very little can be done to prevent damage when the ground ruptures beneath a building. But a great deal of damage resulted not from ruptures, but from shaking alone—and this is the aspect of the San Fernando quake that has some engineers worried the most.

Total destruction may exceed the actual earthquake damage. Above, the 1933 shock knocked down the first floor wall, which in turn caused the second story to fail and sag downward, pulling the roof along with it.

Cummings Valley School was destroyed by the 1952 Kern County earthquake. Made of virtually unreinforced concrete walls and wood roof, it has been called "a classic in poor design, poor material and poor workmanship."

HOW AND WHY EARTHQUAKES CAUSE DAMAGE   41

How to Recognize Fault Features

CALIFORNIA'S SAN ANDREAS FAULT is the most conspicuous rift of its kind in the world. Throughout its 700-mile course from the Mexican border to the Mendocino coast, an almost continuous chain of topographic features clearly marks the fault's path. This unique series of characteristics cuts a definite rift zone across deserts, high mountain ranges, coastal hills, and inland valleys.

The details of the fault can be easily followed for hundreds of miles. But even more important, the fault can be seen indirectly in the great mountain ranges and deep valleys along its route. Above all else, the San Andreas Fault has created much of California's scenic beauty.

Although evidence of the San Andreas Fault is very common, it seldom is recognized by either residents or visitors.

The early morning or late evening air traveler between Los Angeles and San Francisco may be fascinated by the giant shadowy scar that stands out against the Coast Range, but he dismisses it as a trick of the weather.

The rancher in the hot coastal ranges who unexpectedly finds a deep ravine and—of all things—a small pond right on top of an arid mountain ridge may wonder only briefly about cause and effect, but he quickly dams the pond and drives his stock to the welcome water.

The Southern Californian who enjoys his high mountains and flat, hot deserts within minutes of each other probably never considers just how this could happen, nor does he relate their existence with the sharp midnight shudder that rattles the windows and sets the neighborhood dogs to barking.

Prospective home builders often eye with envy the high bluffs and steep cliffs that jut up unexpectedly from normally flat valleys, but they rarely ask how such unusual ridges are caused.

ONCE YOU LEARN TO RECOGNIZE THE BASIC FEATURES OF THE SAN ANDREAS FAULT, YOU CAN FOLLOW IT FOR MILES ALONG PUBLIC ROADS. THE CASUAL HIKER WILL FIND MANY FASCINATING SUBJECTS FOR DETAILED STUDY.

One of the best places to explore the San Andreas Fault is in Carrizo Plain, where recent fault movement has torn a straight gash through the flat grasslands. From the air, the fault stands out as a giant scar; from ground level, the jagged stream beds and bold scarps present a clear record of violent movement.

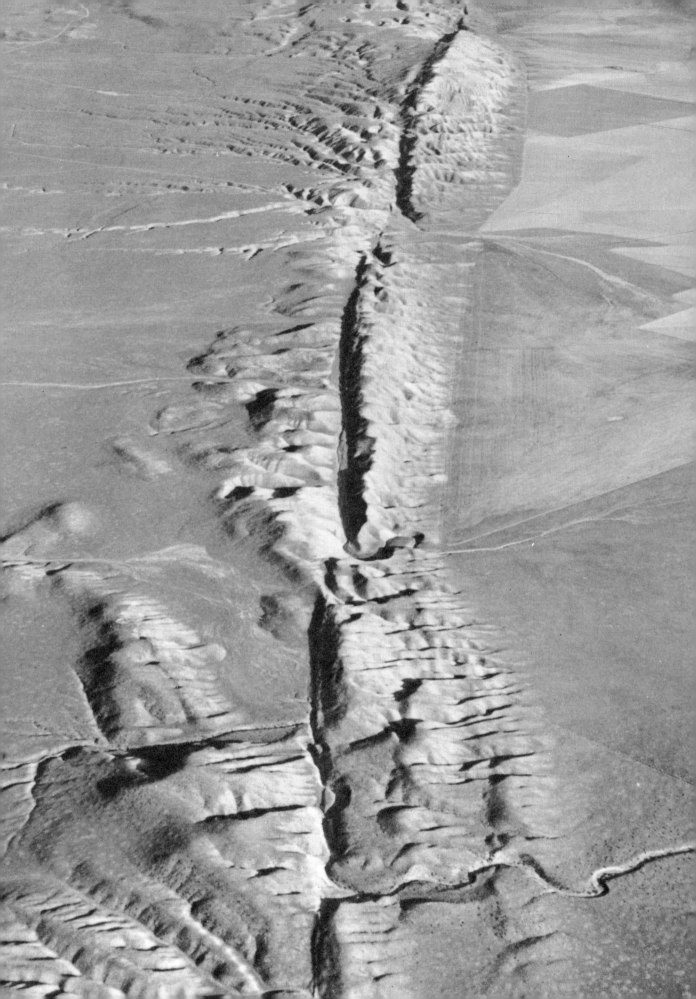

Sailors and fishermen delight in the protective bluffs and straight lines of Tomales and Bodega Bays, while camera bugs dig their tripods into the steep banks to catch the flight of wing and sail across the calm waters, unmindful of the great upheavals that created this gentle land.

The major inland highway routes that link Los Angeles and central California all pass through mountain gaps that were at least partially formed by the San Andreas Fault. Yet no signs mark the fault's path, nor do the hurrying motorists stop to wonder at the abnormally deep cuts through the mountains.

But once you are aware of the fault and its general location, the landscape takes on a new dimension. What might normally be passed off as just another pond of water or mountain suddenly becomes an important sign of faulting that lays bare the route of recent movements.

Active faults such as the San Andreas are notoriously inconsistent in topographic expression. Steep escarpments may reverse direction within a few feet, then suddenly die out to be replaced by a trench or small pond—or nothing at all. But there are two general characteristics of the San Andreas that aid immeasurably in following its course.

First, the fault is essentially a linear feature, so the surface details tend to be arranged in a straight line. When the fault momentarily becomes lost from sight under a landslide or sand dune, you will soon learn how to find its extension in the next valley.

Second, many of the fault features disrupt the normal landscape and appear in direct contrast to what we would normally expect to find. This is particularly true of ridges and valleys that follow the fault rather than the normal high-to-low drainage patterns. And almost nowhere else except along the fault zone will you find a steep gulley cut high in the side, or actually on top, of a ridge.

Since tracing the fault depends on the recognition of topographic features that are not always obvious, the best time for exploration is during morning or late afternoon hours when the light is low and the landscape stands out in sharp relief.

RECENT ACTIVITY

While all of the very large, eroded escarpments that stand thousands of feet high along the edges of the fault zone are obviously the result of millions of years of faulting, most of the lesser details you will find are along the line of most recent activity within the zone. This activity may have occurred hundreds or thousands of years ago, but it is still "recent" geologically. And in most cases, the features are the result of several huge earthquakes spread over any number of centuries. Many fault movements are usually required to create a new ridge or depress a block of land until it becomes a sag pond, and earthquakes large enough to rupture the surface are few and far between.

It is true that very recent large earthquakes such as those of 1906 and 1857 "freshened" the fault features and in some cases created new low escarpments and sag areas that can still be seen today. But these are comparatively rare, and it is very seldom that any significant feature can be definitely assigned to a single movement.

The expression "line of most recent activity" shows up repeatedly in the following chapters as a reminder that movement along the San Andreas Fault has been going on for eons, but that we can see only the work of a comparatively short time.

In the next 5 pages, each fault feature that you will find along the San Andreas is explained separately. In reality, however, the features are intermixed in more significant arrangements that make the fault easier to find and to follow.

FAULT VALLEY

A fault valley is the principal topographic expression of the San Andreas Fault. These valleys cut into mountains and lowlands alike. Many are narrow enough to stand out from the surrounding country (Lone Pine Canyon, Cuddy Valley, and the area west of Manchester), but at times it may be miles wide and cover all of the area between dominant mountain ranges (south of Hollister).

The valley often looks like a giant "rift," and the term "San Andreas Rift Zone" often appears on maps and "rift valley" is sometimes used in describing the Fault.

The valley along the San Andreas is due primarily to rapid erosion that takes place in the crushed and broken rock within the fault zone. While the rock formations on either side remain solid, the rubble caused by the shearing action along the fault is easily weathered and eroded, washed away by rain and rivers that naturally tend to push their courses through the soft rock.

SADDLES

Where the fault line cuts across a steep slope or ridge top, the resulting zone of crushed rock is quickly

San Gorgonio Pass is a narrow block that was "left behind" when movements along active branches of the San Andreas fault zone elevated the mountain ranges on either side.

eroded into a saddle or shallow swale that interrupts the normal ridge contours. Good examples of this are at Big Pines Summit, on a slope west of Gorman, and in Marin County.

SUNKEN BLOCKS

Similar in appearance to a fault valley, but formed by a different process, are the many sunken blocks (grabens) along the fault. When two parallel branches of the fault system are active at the same time, the block of land between them often is pushed down by the movement, or is left behind as the mountains rise on either side. This results in the formation of a broad, flat valley with steep slopes on either side. Once the basic structure is established, then erosion often adds its work to accent and emphasize the sunken block.

Both the Cholame and Imperial Valleys were formed along the San Andreas Fault in this way, and Owens Valley represents a huge sunken block along the Sierra Nevada fault system.

NARROW TROUGHS AND RIDGES

The multiple fracturing that may go on within a fault valley often results in several crustal blocks rising and falling in an irregular pattern. On the surface, this action is expressed by a series of narrow, long ridges separated by deep troughs or gullies. This type of development can also be created or emphasized by differential erosion along slices of the fault that have been broken up by sharp movement.

Alternating ridges and depressions are particularly noticeable on the San Francisco Peninsula and North Bay areas. The San Francisco Water Reservoirs at Crystal Springs and San Andreas Lakes are located in just such an area of mixed uplift and depression.

SCARPS

A scarp is a steep cliff or ridge that is formed by sudden earth movements—usually vertical—along fault lines. This is by far the most common form of evidence

Rounded, elongate hill in Carrizo Plain is typical of a scarp that has been blunted by erosion. Most recent fault movement has been along the base of the steep side.

along the San Andreas Fault and in some instances is the only evidence of recent rupture. The highest and oldest scarps are thousands of feet high, such as the sheer slopes of the San Bernardino and San Jacinto Mountains. Such high escarpments are the result of repeated fault movements, each of which contributed a few feet to the uplift. The upper, older parts of these formations are greatly eroded.

But scarps may also be formed by horizontal movement. In an area of mixed high and low relief, a horizontal movement may very well break open a hill or ridge and expose a steep interior face along the line of rupture. (For a simple demonstration of this phenomenon, place a large sheet of paper flat on a table. Put both hands palm down on the paper, then move one hand toward your body and one away from it to approximate horizontal movement. Notice how the paper folds into angular "hills." If the paper could be cut or torn down the middle—much as a fault cuts the earth's surface—these hills would be separated into half-domes.)

There are hundreds of scarps along the San Andreas Fault, ranging in height from a few feet to a mile or more. Erosion may wear away the sharp features of a weak scarp, leaving a rounded hillock, but usually the unique shapes are still easy to see. Of course, only one face of a scarp is steep—the other is a gradual slope. Scarps that have their steep face toward the

mountains usually are better preserved, since erosion tends to emphasize them.

STEEP MOUNTAIN FRONTS

Where mountain ranges are bounded by fault lines, the slopes that are in direct contact with a fault tend to be very steep, without any foothills. Such mountains owe their existence to upthrust along a fault line rather than to any gradual process.

The San Jacinto Mountains near Soboba Hot Springs, the southern slopes of the San Bernardino Mountains at City Creek, and the eastern face of the Gabilan Range south of Hollister illustrate this fault phenomenon.

SIDE HILL RIDGES

Side hill ridges are the result of surface rupture and erosion. When an earthquake breaks the surface along the side of a ridge, the line of weakness formed invites erosion. The downhill side of the break is still resistant, however, and it stands firm while the crushed rocks along the fault line are steadily worn away. Eventually, the edge of the downhill slope juts above grade and forms a secondary ridge that has nothing to do with the normal forces of mountain building but is primarily an erosional feature of faulting. There are several side hill ridges of this type along the San

Fault movement caused the 1906 earthquake to create several low scarps such as this one in Marin County. Scarps of this size were also common in the 1940 Imperial Valley and 1952 Kern County earthquakes.

Andreas Fault; good examples can be found on Anzar Road west of San Juan Bautista and Cienega Road south of Hollister.

SLICE RIDGES

Faulting can create new ridges in two other ways. Most often, rock is squeezed between two parallel faults and forced upward into a hogback or long slender ridge. These are quite common along the San Andreas Fault. The rift valley west of Palmdale contains a very prominent example.

Occasionally a particularly severe earthquake or chain of movements may cause a "slice" of rock or mountain slope to break away from its parent mass and begin to move along within the fault zone. With each subsequent earthquake, the slice is moved farther away from its original mooring until it seems to float free. Eventually, the ridge is either broken into a mass of rubble by the continued shearing actions or is eroded away as it wallows in the fault valley. A large slice ridge of the latter type can be seen east of Cienega Road south of Hollister.

Far more common than slice ridges are isolated rocks that are set adrift by fault action. These rocks are broken away either from the beds along the edge of the fault zone at the surface or from the bedrock areas far below the surface. In either case, the rocks become isolated and are moved around at random by shearing action within the fault zone. After they have been forced to the surface, erosion eats away the surrounding crushed rock and leaves them standing in unexpected places. The shores of San Andreas and Crystal Springs Lakes are marked by several isolated rocks, and isolated exposures of the Punchbowl sandstones can be seen within the fault zone near the eastern end of Lone Pine Canyon. The hills directly above Gorman have several lava outcroppings that are strangers to the land.

JUXTAPOSITION OF DIFFERENT ROCK TYPES

Just as earthquakes can offset fences and tree lines, they can also shift rock formations. Thus, for much of its course, the San Andreas Fault marks the dividing line between radically different rock types. This juxtapositioning of rock formations provides geological clues to the amount of movement along the fault (map on page 23). Since this is essentially a scientific phase of fault determination, only rock hounds will be able to differentiate between most of the various deposits along the fault. But everyone can see the startling differences in color and rock patterns at Whitewater Canyon, Devil's Punchbowl, Vincent Gap, and Mill Creek north of Beaumont.

This startling contrast in rock types is found in Pacoima Canyon along the San Gabriel Fault, a dead branch of the San Andreas system. Long-term faulting has moved granodiorite (left) into face-to-face contact with gneiss (right).

C. W. Jennings

LANDSLIDES

The California Coast Ranges are known for their abundance of landslides, and one contributing factor is the San Andreas Fault. The fault causes landslides in two ways. First, the steep mountain slopes associated with fault-controlled ranges are ideal ramps for landslides. Second, the multiple fractures that occur within the fault zone shatter and weaken the rocks in such a way that they are ready for a landslide the first time they become overloaded with water.

Repeated fracturing of rocks along the steep northern face of the San Gabriel Mountains aided in setting the scene for the historic 1941 mudflow at Wrightwood. The eastern slope of Slack Canyon north of Parkfield is one continuous landslide area owing to the proximity of the fault and to the rock types in the fault zone. And at Mussel Rock, where the San

Andreas Fault goes out to sea, all traces of the fault line are lost in a giant landslide area.

CRUSHED AND DEFORMED ROCKS

When two hard faces along a fault line rub together during an earthquake, they crush and grind rocks between them. The rocks become physically altered by the massive mechanical forces, and they also undergo extensive chemical alteration that converts many of the minerals to clay. Fault zones thereby acquire a buffer of this crushed rock. Such a shear zone may be only as wide as a pencil in a small fault, or it may be hundreds of feet wide in an active and complex fault zone such as the San Andreas. Frequently, repeated movements reduce rock to a clay-like substance, often red in color, called gouge.

OFFSET STREAMS

One of the best ways to determine the line of most recent activity within the San Andreas Fault zone is by studying the routes of rivers and streams that cross the fault at right angles. Since displacement is accumulated slowly by repeated movement, streams tend to hold onto old channels, even though they become increasingly offset. Eventually, the stream bed assumes a peculiarly curved shape that is unique to fault topography. Carrizo Plain is the best place in the world to see offset streams—almost all the drainage lines show displacement. More good examples can be seen near Valyermo, south of Cholame, and at most other sections of the fault where movement has been consistent and where streams cross the fault rather than flowing along it.

SAG PONDS

A sag pond is an enclosed, depressed area within a fault zone. It is caused by multiple fracturing, uplifting, and tilting of the earth so that all drainage avenues are cut off. The undrained reservoirs thus formed then become logical basins for runoff from surrounding high ground. Sag ponds are almost as common as scarps along the San Andreas Fault, and they show up at all elevations and in all types of terrain. Among the most conspicuous are Lost Lake at Cajon Pass and the Elizabeth Lake basin. Hundreds of sag ponds are visible along the fault, and many have been modified into small reservoirs or watering holes for livestock. It is a simple job to convert a sag area into a reservoir, since two or more of the sides are formed by scarps and similar fault features.

Offset stream channels in Carrizo Plain are marked by a great consistency. Small, intermittent movements allow the streams to hold onto same channels even though displacement may total hundreds of yards.

IMPOUNDED GROUND WATER

When the earth shifts along a fault line, subterranean water channels and gravel beds may become disrupted and offset to such a degree that water cannot flow its usual route to other ground. Even if the channels are not completely blocked, the crushed rock within the fault zone may turn to clay (gouge) and form an impenetrable barrier. Under either of these conditions there is a back-up of water on the uphill side of the fault and a relatively arid zone on the downhill side.

If the ground water thus impounded has enough pressure behind it to force a path to the surface, a spring or small lake is formed. If the water doesn't find its way to the surface, the water table may be raised high enough to support an uncommon concentration of vegetation. In Southern California's desert areas, lines of impounded ground water clearly mark the line of faulting across sand and rock. North of Indio, a long series of oases and scattered palms rests on the fault. In Whitewater Canyon and on the southern extension of Big Rock Creek, dense lines of vegetation spring from otherwise barren flat lands. In Seven Palms Valley, a wide strip of palms and brush cuts through the hot desert floor.

HOT SPRINGS

Hot springs are common along the San Andreas Fault, the San Jacinto Fault, and other major faults in Southern California. Many have been commercially developed into popular health and resort areas. The heat probably is caused by friction of the rocks slicing past one another within the fault zone.

This very young stream bed has been offset about 45 feet by repeated horizontal movements along the San Andreas Fault during the past 250-300 years. Slopes above the stream bed have been cleanly sliced by the movements.

Exploring the Southern San Andreas Fault

IN SOUTHERN CALIFORNIA, THE SAN ANDREAS FAULT ZONE IS WIDE AND COMPLEX. BUT TRACES OF RECENT MOVEMENT ARE EASY TO FIND IN THE ARID DESERT AND ALONG RUGGED MOUNTAIN CRESTS.

IN SOUTHERN CALIFORNIA the San Andreas Fault system extends from the Mexican border to the Transverse Mountain Ranges west of Tejon Pass. In many respects, the fault in this area is quite different from its central and northern sections.

First and most important, there is no single San Andreas Fault line that is positively traceable for the entire distance. Near the Mexican border, very little is known about the fault, even though it is active. The faults in Mexico and the Gulf of California that are considered the southern extensions of the San Andreas system disappear under the alluvial deposits of sand and gravel that cover the lower part of California. The 1940 earthquake near the border provided one of the few real clues to the nature of the fault in this area.

East of the Salton Sea, one branch of the San Andreas system becomes traceable near Salt Creek, where several obscure branches gradually come together into a fairly definite belt. This fault line extends north along the Mecca and Indio Hills and then breaks into the Mission Creek and Banning Faults that in turn extend to San Gorgonio Pass.

West of the Salton Basin is the San Jacinto Fault, a long-active branch that many geologists believe is currently the main branch of the San Andreas system. It is traceable along the southwestern edge of the Santa Rosa Mountains, through the Anza-Borrego Desert area, past the towns of Hemet and San Jacinto, then between Colton and San Bernardino, and northwest along Lytle Creek Canyon and Vincent Gap to the Devil's Punchbowl area. This readily identifiable line, plus the frequent earthquake activity along its route during the past century, provides strong evidence for claims that the San Jacinto Fault is now far more important than the trace of the San Andreas east of the Salton Sea.

In the San Gorgonio Pass area, the eastern trace of the San Andreas zone becomes even more complex. The Banning Fault forms the major structural break on the north side of the Pass, but then apparently disappears in the slopes above Beaumont. The Mission Creek Fault passes through Desert Hot Springs, intersects Morongo Valley Canyon, and then extends west along the south fork of Whitewater River and south to San Gorgonio Canyon near Banning. The Mill

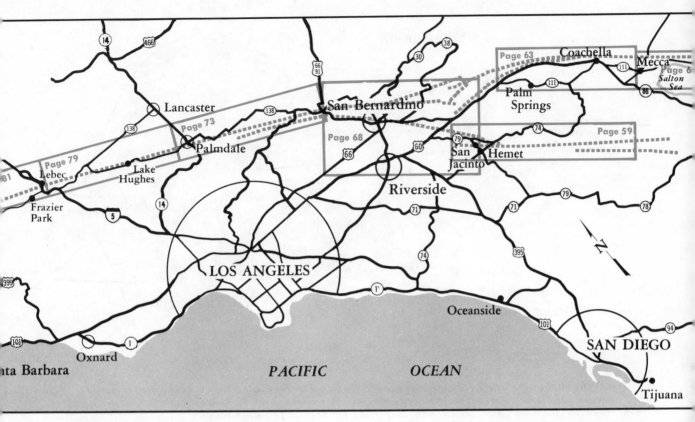

Map shows the area covered in this chapter. The page numbers within the rectangles refer to maps for the subchapters where the territory inside the boxes is described.

Creek Fault appears almost as a westerly extension of the Mission Creek Fault (the two may be connected), as it starts along the North Fork of Whitewater River and takes a westerly course toward Cajon Pass.

What is usually referred to as *the* San Andreas Fault in this area assumes definite identity at Burro Flat north of Banning, cuts northwest to Pine Bench and the northern edge of San Bernardino, merges with the Mill Creek Fault at Waterman Canyon, and then continues in an unbroken line to San Francisco.

Between Cajon Pass and Palmdale the San Andreas cuts a straight line through the northern slopes of the San Gabriel Mountains. It is marked by an almost continuous chain of scarps, sag ponds, and troughs. The San Jacinto Fault parallels the San Andreas for much of this distance and can be clearly seen in the area between Highway 2 and the Devil's Punchbowl.

Between Palmdale and Tejon Pass the San Andreas Fault passes through Leona Valley, the Lake Hughes area, Pine and Oakdale Canyons, and then parallels State Highway 138 to Gorman. West of Tejon Pass the fault zone is traceable through the Frazier Park area, Cuddy Valley, and the San Emigdio and Santiago Creek Canyons.

Another significant feature of the San Andreas Fault zone in Southern California is its angular character. In general, the fault is known for its strict linearity, and in Central and Northern California it rarely swerves from a straight and narrow path. But in the South the fault makes a general shift in direction near San Gorgonio Pass, and then makes an even sharper bend from almost due west to northwest again in the area between Tejon Pass and State Highway 33. The degree of change at San Gorgonio Pass and the reasons behind it are clouded by the great complexity of faulting and the many identifiable fault lines that converge on the area. The pronounced curve west of Tejon Pass, however, is much easier to trace and gauge, and therefore has been the subject of quite a bit of speculation.

There are two factors that may have contributed to the change in direction. First, the San Andreas Fault becomes involved with the complex western end of California's Transverse Mountain Ranges. The east-west trend of these mountains is certainly at odds with the northwest-southeast trend of the fault system, and the great jumble of conflicting pressures may have forced the San Andreas to alter its course and find weaker lines of resistance.

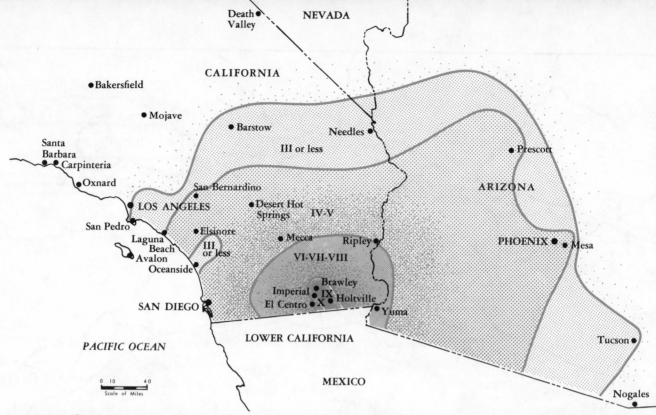

Among the most recent significant earthquakes along the San Andreas Fault system were the 1940 Imperial Valley and 1948 Desert Hot Springs jolts. The isoseismal map above shows the area of Southern California that felt the 1940 quake, which originated on the Imperial Fault branch. The magnitude was 7.1.

Second, the San Andreas Fault meets the Garlock Fault not far from Frazier Park. Although the Garlock apparently is not as old or as strong as the San Andreas, it nevertheless is capable of exerting great force. At one time the San Andreas and the Garlock may have struggled for dominance, and the conflict could have forced a curve in the San Andreas Fault line. The current theory that the Garlock and Big Pine Faults are actually offset branches of the same original break adds support to the idea that there has been interaction between the Garlock and San Andreas systems.

Regardless of the reason behind it, this bend in the San Andreas undoubtedly is important to the state's geologic history. Some geologists, in fact, blame it for the thrust faulting responsible for the 1971 San Fernando earthquake (see page 132).

Fault features are generally well exposed throughout the Southern section. Roads follow along the fault line for several miles.

PRINCIPAL EARTHQUAKES

Earthquake of January 9, 1857—Tejon Pass
Est. Magnitude: At least 7¾; Maximum intensity: X.

This was the strongest earthquake to hit Southern California during recorded history. Even though population was sparse in the areas of greatest intensity, the few first-hand reports that are available indicate that the shock was probably as large as the 1906 earthquake in Northern California.

The earthquake was centered near Tejon Pass, and the surface along the San Andreas Fault was probably ruptured from the San Bernardino Valley to Cholame Valley, a distance of more than 225 miles. The shaking was most intense at Fort Tejon, where every building was cracked or knocked down. It was severe in Los Angeles, San Francisco, and Sacramento. Despite its great magnitude, the earthquake caused only one fatality. In the mountains near Fort Tejon a woman was killed in the collapse of an adobe building.

Most of the information on this earthquake is available only from personal letters, military reports, and scattered newspaper stories. The most detailed report of the shock appeared in a historical summary published in a Visalia newspaper in 1876 that said it was the "heaviest earthquake shock which has ever been experienced in (San Joaquin) Valley. Houses and trees vibrated violently. The solid earth seemed to have lost its stability and a wave-like motion was experienced as on shipboard . . . for a moment nature seemed filled with terror. The line of disturbing force

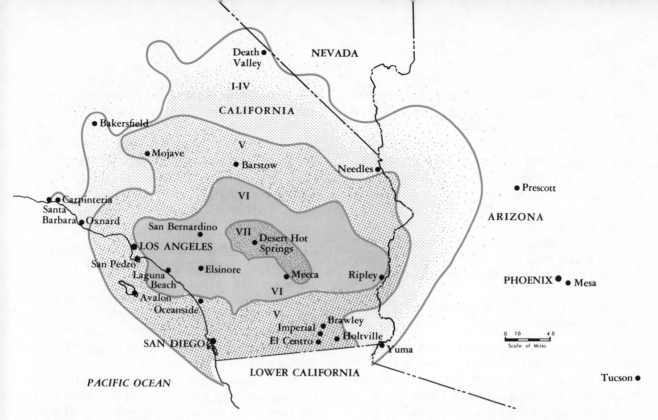

The 1948 earthquake affected an area almost as large as that of the 1940 shock, but intensities were considerably less. Magnitude was 6.5. Irregular limits of the isoseismal zones are caused by differences in geologic structure that have a great effect on intensity.

followed the Coast Range some 70 miles west of Visalia and thence (southeast) out into the Colorado Desert. This line was marked by a fracture of the earth's surface, continuing in one uniform direction for a distance of some 200 miles."

Letters from persons who were in Los Angeles at the time of the earthquake tell how the tremors built up slowly in the city. Houses rocked and toppled, and people were thrown down in the streets; but because the movement was comparatively slow at this great distance, swaying buildings had a chance to recover slightly between each successive tremor. "If the motion had been short and sudden," wrote one man, "the damage would have been appalling."

Another personal letter, this one printed in a San Francisco newspaper, included a description of surface fissures several feet wide in an area 20 miles south of Los Angeles, and of new streams of water that were created at Paredes. There were several other reports of changes in water levels and stream courses at widely scattered points. The waters of the Mokelumne River were thrown over the banks, leaving the river bed temporarily bare. The current of the Kern River was turned upstream and the water ran over the banks in four-foot waves. The water of Tulare Lake was thrown up on its shores, and the Los Angeles River was thrown out of its bed.

At San Fernando a man reported that "the earth was in fearful agitation, with undulations so quick and rapid as to make it almost impossible to stand. The sensation was very much like that felt on the deck of a small vessel in a heavy chopped sea."

Along the fault line the rupture created a number of scarps and trenches. Some reports indicated that ridges and mole track effects were as much as 10 feet wide. Surface features were particularly evident in the area around Elizabeth Lake. At Gorman an adobe house was shaken down and the road was cut by several parallel breaks.

Near Fort Tejon a miner was spending the night camping out. He was lying awake, when at approximately 8:30 A.M. the earthquake began to shake the ground beneath him. Frightened, he leaped to his feet just in time to see the ground open and swallow his camping gear. There were reports of cows being thrown down steep hillsides by the shock, and one poor beast was reportedly entombed in a large fissure.

According to the few reports available, the earthquake was very severe in Carrizo Plain (see the chapter on Central Section), and probably caused many of the fault features that can be seen in the area today. The lines of displacement reportedly covered an area nearly a mile wide; at one point, a round corral was distorted into an S-shape by the horizontal movement.

The most northerly report of earthquake rupture came from Cholame Valley, where a settler watched while his cabin was shaken down.

A U. S. Coast Survey party doing field work at Santa Barbara found long cracks in the bed of the Santa Clara River, some of them six to eight inches wide. Families living near the river said that six-foot jets of water erupted from the cracks, and that large blocks of earth sank several feet.

Reports of damage in San Francisco generally were concerned with items falling off store shelves, pendulum clocks that stopped, and a few shaken buildings.

Earthquake of July 22, 1899—Cajon Pass
Maximum intensities: VIII-IX.

The greatest intensity of this major earthquake was in the San Bernardino Mountains near Cajon Pass. The road in Lytle Creek Canyon was blocked by landslides in many places, and Cajon Pass Road was covered with debris for a distance of 2,000 feet. The greatest property damage was in San Bernardino, Highland, and Patton. Minor damage to residences was reported in Riverside, Redlands, Pomona, Pasadena, and Los Angeles.

A San Bernardino newspaper reported this unfortunate sequence: "A man named Baker was being shaved in a Third Street barber shop at the time of the early afternoon shock. He was badly frightened and rising suddenly from the chair, his face came in contact with the razor, the result being a long deep gash. While the surgeon was putting in the stitches, a second shock was felt. Baker made another jump and was painfully prodded by the needle."

Earthquake of December 25, 1899—San Jacinto
Maximum intensities: IX-X.

This shock was centered on the San Jacinto Fault near San Jacinto and Hemet. Brick and adobe structures were shaken down in both towns. Six Indians were killed and eight were injured by collapsing adobe walls at Soboba. The shock was heavy at Riverside and was noticed as far away as San Diego. There are several large sinks in the mountains southeast of San Jacinto; these reportedly were created during this earthquake.

Earthquake of September 19, 1907—San Bernardino
Magnitude: 6. Maximum intensities: VII-VIII.

Although heaviest at San Bernardino, this quake was felt from San Diego to Santa Barbara. There

were three distinct shocks at San Bernardino; the shaking at San Jacinto was the heaviest since 1899. Riverside also reported its strongest shock in years, and large buildings swayed in Los Angeles. Many landslides were reported north of San Bernardino.

Earthquake of June 25, 1915—Imperial Valley
Magnitude: 6¼. Maximum intensity: IX.

The epicenter was located near Calexico and El Centro, but it is impossible to say whether the San Jacinto Fault or the Imperial Fault was to blame. There were two distinct and equal shocks, 57 minutes apart. Six fatalities resulted in Mexicali when people fled from buildings after the first shock but decided to return only to be crushed by walls thrown down during the second main tremor. A farmer reportedly was driving a mule along a path near the fault at the time of the first shudder, and the beast was knocked off his feet. An hour later the farmer and mule were returning home along the same route and the poor animal was knocked down again. The farmer thought the spot was bewitched.

Large areas of irrigated farm land sank a few inches to a foot or more. After the shocks, farmers reported that one-third more water was needed for irrigation because of the large cracks in the ground.

Earthquake of April 21, 1918—San Jacinto
Magnitude: 6.8. Maximum intensities: IX-X.

This was another large quake centered on the San Jacinto Fault. There was no loss of life, but damage was considerable in San Jacinto and Hemet—only two structures of any kind were still standing in San Jacinto after the dust settled. Ground cracks (caused by earth slumping rather than by actual rupture) could be seen for 10 miles along the fault, and there were great changes in stream flow and hot springs output. A press report gave this impression: "The building rocked and twisted like a ship at sea. The earthquake started with a vertical bumping movement, then a twisting and rocking motion, with walls creaking and groaning, windows rattling, pictures on the wall rattling. Doors opened and swayed back and forth." The shock was heavy in Redlands, Palm Springs, and Los Angeles.

Earthquake of October 22, 1916—Tejon Pass
Magnitude: 6. Maximum intensity: VII.

The epicenter was located near the summit of Tejon Pass. Five shocks were felt at Gorman, where grocer-

The 1940 earthquake force threatened to throw down this brick column in Brawley, but weight of the roof provided just enough pressure to hold it in place.

Surface rupture of the 1940 earthquake in Imperial Valley revealed for the first time the exact route of the Imperial Fault, a branch of the San Andreas system.

ies were thrown from store shelves and dishes fell from cupboards. A crack an inch wide and several hundred feet long opened in the concrete highway. A report from Ozena (28 miles southwest of Tejon Pass) said the shock was preceded by a sound "like an automobile coming with the engine working at its best. When the sound reached me it felt as if it came under the ground and under the table at which I sat. It sounded like thunder under the ground as it went by." In Cuddy Valley, Mr. Cuddy had his car mounted on four jacks and the shaking toppled it off.

Reports of minor shaking came from as far as San Diego and Fresno. Near Elizabeth Lake two shocks were felt and a man was nearly thrown down by the tremors.

Earthquake of May 18, 1940—Imperial Valley Magnitude: 7.1. Maximum intensity: X.

This is the earthquake that exposed the exact line of the Imperial Fault, which is the only known section of the San Andreas system near the U. S.-Mexico border. The ground was ruptured for 40 miles from Volcano Lake in Baja California to a point near Imperial. There were seven deaths, and property loss was in excess of $5 million. The epicenter was located east of El Centro. Eighty per cent of the buildings at Imperial were destroyed; 50 per cent of Brawley's structures were severely shaken. Indirect damage to crops was substantial, owing to the disruption of drainage and flooding immediately after the break. Horizontal displacement across the completed but as yet unfilled International Canal was 14 feet, 10 inches, creating a permanent change in the U. S. - Mexico boundary line. The Alamo Canal in Baja California also was offset, and water spilling out of the broken channel caused a local flood.

Perhaps the most conspicuous area of surface rupture was on State Highway 98, eight miles east of Calexico. The roadway was broken by a four-foot scarp, and rows of trees in an orange grove south of the highway and west of the Alamo River bridge were offset almost 10 feet. The offset is clearly shown in the aerial photo on page 20; unfortunately, the trees were torn out in 1964, so no traces remain of this dramatic movement. The maximum horizontal dis-

Outside wall on this old brick building in Holtville failed to withstand the 1940 earthquake, and toppled onto a car parked below. The building's interior partitions were scarcely damaged.

placements of the earthquake—approximately 19 feet —were measured in the area just south of the orange grove. Vertical movements were very slight.

The main shock took place at 8:36 P.M. A large aftershock at 9:53 P.M. did most of the damage at Brawley. There were dozens of other aftershocks during the next four days. Loss of life was very low in relation to structural failure; however, if the earthquake had occurred during the daylight hours when the stores of Imperial and Brawley were filled with shoppers, the number of deaths undoubtedly would have been much higher.

Reports from residents in the affected area indicate that the main shock lasted only 10 to 15 seconds.

The earthquake did considerable damage in Mexico and Arizona. The Yuma Weather Bureau reported that "in the lower Yuma Valley, cracks opened up in fields and water and sand gushed up; wells spouted as much as 15 feet into the air; ground levels were so affected that some fields will have to be releveled."

One of the few places where the 1940 offset can still be seen is along Old Highway 80 (Evan Hewes Highway) about 5 miles east of El Centro. From the James Road intersection, look west along the line of telephone poles just north of the highway; a 32-inch lateral offset can clearly be noted between the third and fourth poles. A distinct crack in the highway can

also be seen opposite this point (see photos on opposite page).

Earthquake of December 4, 1948—Desert Hot Spr. Magnitude: 6.5. Maximum intensity: VII.

The epicenter of this fairly strong earthquake was in the unpopulated mountains east of Desert Hot Springs. The towns of Twentynine Palms and Indio suffered the most damage, but neither was seriously shaken. Broken windows and a few slight building cracks were the most serious results. A bank vault was sprung at Twentynine Palms. An isoseismal map showing intensities is on page 53.

Earthquake of March 4, 1966—Imperial Magnitude: 3.6. Maximum intensity: V.

In terms of magnitude and intensity, this was a small earthquake, and it caused virtually no damage. Its significance lies in the fact that surface ruptures and slight horizontal displacements accompanied the shake. Until this time, it was thought that a much larger magnitude was required to cause a surface rupture along the fault line.

The maximum surface displacement occurred at the very point on Highway 80 just west of the James Road intersection where the 1940 offset took place (see photos above on opposite page).

C. R. Allen

These two photos were taken at the same location on Highway 80 east of El Centro. LEFT: Rotated concrete slab shows the 18-inch displacement caused by the Imperial Valley earthquake of 1940. Line of telephone poles was displaced an equal distance. RIGHT: This photo was taken after the 1966 Imperial quake. Highway has been resurfaced, but arrow shows location of crack along old white line. Total displacement is now 33 inches— the 15 inches additional were caused by creep or small jumps during small earthquakes.

Earthquake of April 9, 1968—Borrego Mountain
Magnitude: 6.5. Maximum intensity: VII.

This quake was felt throughout most of Southern California and adjacent areas. Because the epicentral area west of the Salton Sea is virtually undeveloped, there was very little damage. The earth's surface was broken for 30 miles along the Coyote Creek Fault (a branch of the San Jacinto system), with a maximum horizontal displacement of 16 inches about three miles northwest of Ocotillo Wells.

The fault crossed Highway 78 at the crest of the low hill a few hundred feet east of the Burro Bend Cafe at Ocotillo Wells and passed through the middle of the airstrip (dry lake) north of the highway. The only visible remaining evidence of the 1968 displacement is a series of deep fissures in the desert floor southeast of Ocotillo Wells. Although they are gradually filling with windblown sand, these deep fissures are still visible where they cross the old gypsum mine road (about 5 miles east of Ocotillo Wells), 1.9 miles south of Highway 78.

This quake also triggered small displacements on the Imperial Valley Fault, the Superstition Mountains Fault, and the San Andreas Fault near the Salton Sea. This was the first time that movement on one fault had been observed to cause visible movement on other faults.

Malcolm Clark

Fissures near gypsum mine road southeast of Ocotillo Wells opened during and after the 1968 Borrego Mountain quake. They are gradually filling with windblown sand.

The Santa Rosa Mountains have been upthrust along the San Jacinto Fault to form the northeastern edge of Borrego Valley. One branch of the zone separates the valley floor from the lowest slopes, and other lines of movement are visible at higher levels.

Salton-Imperial Basin

THE SAN JACINTO FAULT IS TRACEABLE WEST OF SALTON BASIN; THE SAN ANDREAS FAULT EMERGES TO THE EAST. BOTH ARE ACTIVE PARTS OF THE SAME GENERAL FAULT ZONE.

EVEN THOUGH THE SAN ANDREAS FAULT is known to be quite active in the Salton-Imperial Basin, it is difficult to define and almost impossible to trace. The 1940 earthquake disclosed part of the system, but between this fault and the Salton Sea area there is no surface evidence of any other significant fault lines.

THE SAN JACINTO FAULT

The San Jacinto Fault lies west of the Salton Basin. At its southern end it has created a regular mountain wall that encloses the northern end of Borrego Valley. Most of the isolated granite hills southeast of Borrego Valley (such as Superstition Mountains) are also controlled by various branches of this fault. No good public roads follow the fault in this region, but the jeep traveler can see traces of recent activity in the northwest-trending segment of Rockhouse Canyon, a half mile southeast of Hidden Spring in Anza-Borrego Desert State Park. Farther north, the hiker can see some excellent exposures of the fault in the

Pacific Air Industries

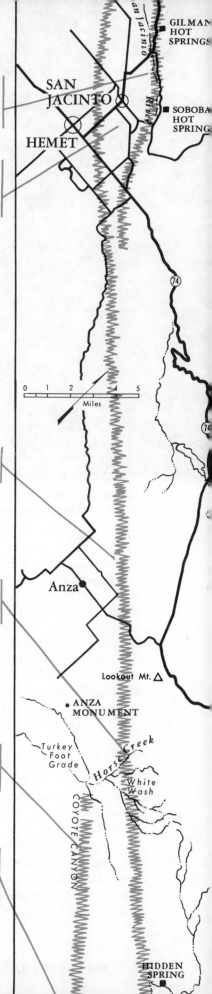

One branch of the fault lies along the steep front of the San Jacinto Mountains. Hot springs were created by fault activity.

San Jacinto and Hemet are close to active branches of the fault and have had more than their share of earth-quake jolts during recorded history.

Northwest of Anza, horizontal fault movement has offset canyons and stream beds.

canyons of Horse Creek and White Wash, reached from the bottom of Turkeyfoot grade southeast of Anza.

On State Highway 79 north of San Jacinto the fault runs along the base of the very steep and straight slopes of the San Jacinto Mountains. Both Soboba and Gilman Hot Springs are located on the fault line and owe their existence to water rising along the fault. Part of State Highway 79 and Soboba Road are located in a fault valley along the San Jacinto River. Vertical displacement along the fault north of Soboba Hot Springs has been more than 1,000 feet.

Farther north, the fault continues to form the topographic borders of the San Jacinto Range and Badlands. It then cuts across the San Timoteo Hills and into the San Bernardino Plain (see page 68).

EAST OF THE SALTON SEA

On the east side of the Salton Sea another fault trace becomes recognizable near Durmid, where sandstone

Good views of fault features are available to hikers in the Horse Creek area.

A well-traveled jeep trail follows the fault from Anza Monument south through Coyote Canyon.

Southern extension of the San Jacinto Fault has created mountain front along the northern end of Borrego Valley (photo left).

This canyon north of Anza has been substantially offset by repeated movements along the San Jacinto Fault, until the lower reaches are now located several hundred feet northwest of the upper section.

and shale beds on the southwest side of the fault have been upended and contorted near the fault. At the Salt Creek bridge the fault line separating these shales and the pink sandstones on the northeast side of the fault is particularly noticeable. The fault line itself is marked by a wide zone of reddish crushed rock (gouge) easily seen from the railroad bridge over Salt Creek.

From the Salton Sea the fault line continues northeast along the edge of the Orocopia Mountains. State Highway 111 parallels the trace on the northern shores of the Salton Sea.

PAINTED CANYON

East of Mecca, the fault crosses the mouth of Painted Canyon, a beautiful exposure of uplifted and tilted sedimentary rocks on the south-western edge of the Mecca Hills.

To reach the Canyon, take Highway 195 northeast from Mecca, then turn north on Painted Canyon Road about one-half mile beyond the Coachella Canal crossing. This graded road follows the power line for about 2 miles, then swerves into the entrance

of Painted Canyon and continues another 3 miles through towering cliffs of rock that have been folded and thrust into weird shapes, then weathered into a colorful array of shadowy caves and bright colors. The road passes an area of highly deformed rocks in the canyon that is associated with ancient branches of the San Andreas system, and ends in a parking area. The most recent trace of the fault is evidenced by the reddish hill on the right as you enter the canyon.

Along the walls of the canyon are steep exposures of crushed and broken granite and gneiss, with some volcanic rocks. Several zones of fault "gouge" can be easily examined, and the upturned unconformities that have resulted from earth forces are very well exposed.

For an excellent view of this red gouge, turn right at the entrance to the canyon and ascend the trough that is carved along the fault line. Note the many blocks of highly sheared and distorted rocks that are included within this red material.

The Mecca Hills are made up of similar formations of normally flat sedimentary beds that have been distorted by the squeezing action along the Fault.

Clarence R. Allen

Rocks in Painted Canyon have been uplifted and tilted by faulting, then weathered into a spectacular exhibit of weird shapes and wild colors.

Fault movements have warped and tilted rock formations along the coast. Evidence of faulting is particularly evident near Salt Creek bridge.

In White Wash, 11 miles southeast of Anza, the San Jacinto Fault exhibits some evidence of thrusting, where older rocks above the fault apparently have moved upward in relation to more recent gravels.

John C. Crowell

Southernmost trace of the fault zone dies out in the desert east of Salton Sea.

A steeply-inclined zone of reddish "gouge" marks a branch of the San Andreas Fault system in the Mecca Hills. The rocks on the right are nearly horizontal conglomerates and sandstones; but rocks left of the fault are badly crushed volcanics and granitics.

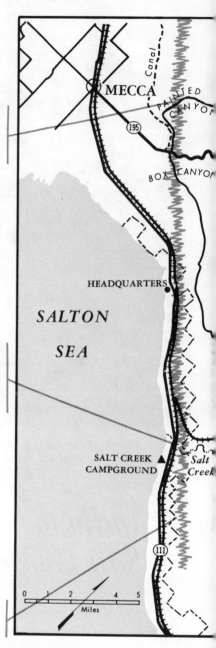

The San Andreas Fault cuts a spectacularly straight gash through the Indio Hills, which are made up of rocks that have been tilted and thrust upward by earth movements.

Indio to San Gorgonio Pass

THE MOST COMPLICATED SECTION OF THE SAN ANDREAS FAULT IN CALIFORNIA IS IN SAN GORGONIO PASS, WHERE NO LESS THAN FOUR BRANCHES MAKE UP AN ACTIVE SYSTEM.

NORTH OF INDIO, the San Andreas Fault system becomes much more recognizable on the surface, but it is still complex. There is no single trace but rather a group of roughly parallel faults that are all considered members of the same family.

Northwest of Indio, the main fault line that is traceable along the edge of the Mecca Hills splits into the Mission Creek and Banning Faults. The Mission Creek branch forms the northeastern edge of the Indio Hills, then veers north past Desert Hot Springs and becomes mixed in with the Pinto Mountain and Mill Creek Faults that converge in the San Bernardino Mountains above White Water.

RED CANYON

For those interested in out-of-the-way explorations, a good place to visit the fault in the Mecca Hills is in Red Canyon. To reach the area, take Airport Boulevard east from Thermal, cross the Coachella Canal, and keep right along the northeast base of the flood control embankment. About 1.8 miles after crossing the canal, you can turn left into **Red Canyon** on a road that is normally passable for family cars except after rainstorms.

After twisting and turning up the canyon for 1.5 miles, you will pass through a wide zone of red gouge and then come out into a surprisingly open valley carved along the fault. The most recent fault trace is along the southwest side of the valley, and you may notice an abrupt foot-high scarplet at the base of the slope. The most recent displacement here was a half-inch horizontal shift triggered by the Borrego Mountain earthquake (see page 57).

INDIO HILLS

The Indio Hills are clearly fault-controlled. They are only 20 miles long and 2 or 3 miles wide, and are bounded on the north and south by major faults. Like the Mecca Hills, the Indios have been squeezed and tilted out of the flat desert by shearing action along the San Andreas system. In many areas of the low hills the sedimentary beds have been turned on end in spectacular exposures. An easily reached example is on Dillon Road east of Indio. A short trip along the power line road will bring you into close range of the bizarre shapes.

Geologically, these rocks are known as the Palm Spring sedimentary beds. Their odd shapes are due to the squeezing along the San Andreas Fault, which lies just to the southwest, and subsequent erosion that has rounded and pitted the layers. The spectacular range of colors is a result of chemical alteration of the rocks, oxidation of the iron content, and wind and sun effects.

BISKRA PALMS

Northeast of Indio the fault lies on the very edge of the Indio Hills, and the damming of ground water on the uphill (north) side has given rise to a very straight line of palm trees and small oases along the bottom slope of the ridge. The main break of the fault is in front of these trees, which could never survive without the high water table.

Biskra Palms is one of the larger oases along the Indio Hills, and it is near here that the San Andreas splits into the Banning and Mission Creek Faults. You can get a good view of Biskra Palms, and the line of other palms along the fault line, from the northern end of Madison Street. West of Biskra Palms, the

You can actually touch the fault in Whitewater Canyon, where the break shows as a distinct dividing line between different rock types. Vegetation on canyon floor is concentrated north of the fault zone.

Scarps of the Mission Creek branch stand out on the desert floor east of Desert Hot Springs. The steep ridges are clearly visible from Dillon Road.

Banning Fault is exposed here as a series of low scarps mixed in with sand dunes, plus a distinct line of vegetation on the north side.

Thousand Palms Highway crosses both branches of the fault zone and Willis Palms marks location of the Banning Fault and Thousand Palms Oasis lies along Mission Creek branch.

Tilted rock beds that form Indio Hills (photo on facing page) are well exposed near Dillon Road.

Adventurous explorers can find conspicuous fault traces in Red Canyon.

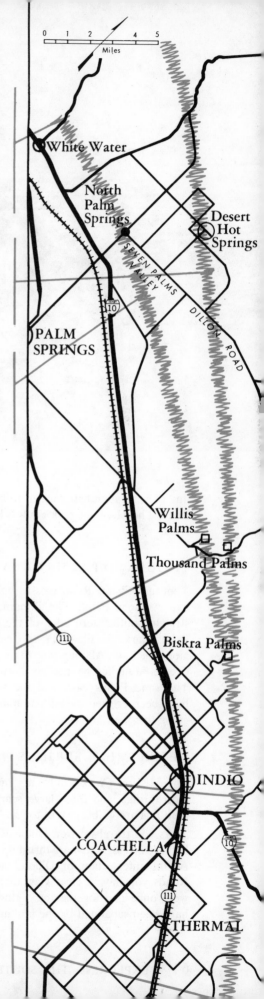

Oases such as Biskra Palms are dependent on the high water table on uphill side of the fault. Rows of trees tend to hug the lowest levels, where roots can reach down into water that has been impounded by fault movement.

line of vegetation can be seen following the Banning Fault. The Mission Creek Fault is hidden in the Indio Hills.

THOUSAND PALMS CANYON

Thousand Palms Road crosses both branches of the San Andreas Fault. The Banning Fault forms the straight southeastern edge of the Indio Hills; Willis Palms marks the site of a spring caused by movement along the fault. About a mile to the north, the Mission Creek Fault lies on the northern side of the hills. Thousand Palms oasis and the spring that feeds it have been created by the damming of ground water along this fault branch.

DESERT HOT SPRINGS

The Mission Creek Fault cuts right through Desert Hot Springs, and three large scarps along the main fault line are visible east of town.

As you travel west on Dillon Road, watch for the high, steep ridges that are thrust up off the desert floor. At the intersection of Dillon and Mountain View Roads you have a clear view of these ridges, which were once tied together as a single escarpment but have been separated by the knifing action of erosion.

The two largest sections lie east of Mountain View Road, and the third is at the outskirts of town.

SEVEN PALMS VALLEY

To the south, the Banning fault can be seen in Seven Palms Valley as a swath of brush and trees that stands out against the bleak sands (an easy place to see the contrast is at the intersection of Palm Drive and Avenue 20). Scarplets and troughs appear here in parallel series rather than in a straight line; recognition of these surface features is somewhat complicated by the unstable nature of the sand dunes, which also generally run parallel to the fault trace.

WHITEWATER CANYON

One of the best spots to see the Banning Fault is in Whitewater Canyon, north of the White Water Post Office. About two miles up the canyon, the fault line cuts across the valley floor, drawing a very distinct line of vegetation with its impounded ground water and then runs in a straight line through the rocks on the western rim. The main road passes close to the western slopes, so the marked difference in rock types is quite evident from the road. North of the fault are the gray granitic rocks; on the south are the brownish

R. C. Frampton

Desert Hot Springs

Seven Palms Valley

The line of the Banning Fault is readily apparent in this aerial photo taken looking southeast. In the foreground is Whitewater Canyon, where the clean break in the vegetation shows where the fault has dammed underground drainage routes to the lowlands. In the middle distance is a discontinuous series of scarps and depressed areas caused by recent activity. In the background is Seven Palms Valley, where the fault again separates vegetation from arid sands.

Willis Palms, northwest of Thousand Palms Highway, lies along the Banning Fault. In the background the fault line is generally traceable along the curving front of Indio Hills. Thousand Palms Highway also crosses the Mission Creek Fault about a mile north of Willis Palms.

gravels. These two different types of rock do not belong face-to-face in this arrangement; they have been moved into this position by fault action. By taking a short, steep hike up the ridge, you can actually touch the fault line.

At the end of Whitewater Canyon Road, near the trout farm, there is a spectacular exposure of conglomerate beds that have been tilted by fault action and then pitted and shaped by wind and water.

SAN GORGONIO PASS

San Gorgonio Pass is an abnormally deep gash between the rugged slopes of the San Gorgonio and San Jacinto Mountains, the two highest peaks in Southern California. The pass owes its existence mainly to the complex fault activity of the area, and appears to be

a narrow fault block that has not been elevated as far as the steep escarpments on either side (the high, eroded scarp on the south is one of the most majestic in California).

Within the pass, details of fault movement are not as prominent as in other sections. Along parts of the fault line, normal rift topography features, such as troughs, sag ponds, and scarps, are completely absent, and no offset streams mark recent movement.

It is difficult to determine which of the many fault branches within the San Gorgonio Pass actually deserves to be called the main line of the San Andreas system. In the over-all history of the fault system, the Mission Creek and Mill Creek may have been more important to the landscape than the fault northwest of Burro Flat that is usually called the San Andreas.

BANNING FAULT

MISSION CREEK FAULT

The Mission Creek Fault is visible as a series of scarps and patches of vegetation east of Desert Hot Springs (foreground). The Banning Fault lies along the straight edge of the dark band of brush to the south.

The Glen Helen Fault, a short branch of the system, is visible near Devore Road.

A large offset in Cable Creek Canyon is visible from the freeway.

At its western end, the Mission Creek Fault splits into two branches.

The Mill Creek Fault sta[rts] here and is traceable west [as] far as Waterman Canyon.

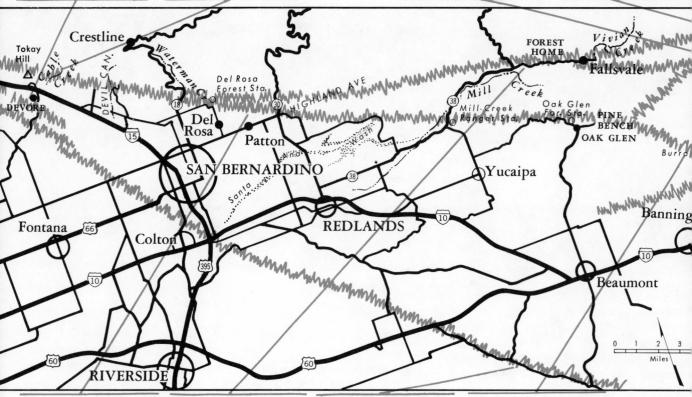

You can see traces of both the Mill Creek and San Andreas Faults as they cross Highway 30 (City Creek Road) within a mile of each other.

Mill Creek Canyon is carved along the Mill Creek Fault. Details are scarce, but the general route of the fault is traceable for several miles.

The San Jacinto Fault passes through San Bernardino Plain and then assumes a north-westerly course parallel to the San Andreas.

The main branch of the fault system—"the" San Andreas Fault—starts at Burro Flat and continues northwest in an unbroken line.

San Bernardino Area

THE FAULT SYSTEM BEGINS TO NARROW DOWN NORTH OF SAN BERNARDINO; THE MILL CREEK AND SAN ANDREAS BRANCHES MERGE, AND THE SAN JACINTO ASSUMES A PARALLEL COURSE.

NORTH OF BEAUMONT, the San Andreas Fault runs northwest along the steep front of the San Bernardino Mountains, skirts the city of San Bernardino, then strikes northwest along Cajon Canyon. For some distance the San Andreas is paralleled by the Mill Creek Fault. Above Beaumont, the two fault lines are about 4 miles apart; the distance between them steadily decreases to the northwest until they merge into a single fault zone at Waterman Canyon.

POTATO CANYON

In Potato Canyon the fault lies along the base of the mountain front to the north, and the mountain slopes are steep and abrupt where they abut the fault. Features of the San Andreas are not as obvious here as they are farther west, but the straight lines of the mountains clearly point to the location of the fault.

MILL CREEK CANYON

Mill Creek Road crosses the San Andreas Fault between the Mill Creek Ranger Station and the mouth of Mill Creek Canyon.

East of Mill Creek Campground, the main road follows the line of the Mill Creek Fault all the way to Fallsvale. The canyon and the road are remarkably straight, in keeping with the linear path of the fault.

Hikers may be interested in the steep Vivian Creek Trail that starts 1 mile east of Fallsvale. It climbs San Corgonio Mountain north of the Mill Creek Fault. By taking short detours off the main trail, you can look down into the green and purple crushed rocks of the fault zone and the giant amphitheaters that have been carved in the rock by action of Mill Creek and a tributary of the Whitewater River. The trail is in San Bernardino National Forest. Inquire locally for details.

As you drive back out of Mill Creek Canyon, watch for differences in rock colors that indicate opposite sides of the fault. To the north the mountains are made up of granitics that show up as white exposures on the slopes; to the south are reddish sandstones, visibly layered and tilted, that are part of a large slice of rock caught between the San Andreas and Mill Creek Faults.

GREENSPOT ROAD

Where Greenspot Road crosses Santa Ana wash, you can clearly see a number of features along the fault. From the top of the flood control embankment just north of the road, you can pick out the difference in rock types across the San Andreas Fault. Looking east across the canyon, the brownish gravels exposed in the cliff are contrasted against gray granitic rocks, which are not vertical but slanted north at about 50 degrees.

From the top of the embankment, you can also see a number of sycamore trees growing along this same line, which reflects the damming of ground water by the fault.

If you turn off Greenspot onto the side road leading to Santa Ana Hydro Plant No. 3, you will cross a low scarp about 200 feet north of the intersection—this is still another extension of the same fault line.

HIGHLAND AVENUE

North of Highland Avenue, a low fault ridge juts up from the normally smooth terrain. This ridge, which has been squeezed up by movements along the San Andreas Fault, is first visible just north of the Church Street-Highland Avenue intersection. The ridge faces the mountains and reverses the normal north-to-south descent of the slope. From Road's End Ranch you can look east along the fault line and see other low elongated hills that have been squeezed by faulting.

About half a mile to the west, Highland Avenue crosses the ridge and descends the steep northern side. If you drive a short distance to the north on one of the side roads, you can look back at the ridge and see how its sharp features stand out on the landscape.

STATE ROUTE 30

This ridge development along the San Andreas can be seen again from State Route 30, just north of the Highland Avenue intersection. The trace of the Mill Creek Fault is also visible one mile north of the intersection. It shows up as a wooded gully that carves a straight gash in the mountains across City Creek to the east. Good examples of the highly deformed sedimentary rocks south of the fault and the gray granites on the north are exposed in the road cuts along City Creek Road.

DEL ROSA RANGER STATION

Del Rosa Ranger Station is located on a low ridge that has been squeezed up between the San Andreas and Mill Creek Faults, which at this point are very close together. The steep side of the ridge faces the mountains, and if you drive to the end of the road behind the main buildings, you can look down into the shallow gully that marks the Mill Creek Fault.

Northwest of the Ranger Station, the two fault lines merge near the mouth of Waterman Canyon. A distinct line of vegetation and trees indicates where ground water has been impounded on the uphill side of the fault zone.

From this point west, the San Andreas Fault cleanly separates two different rock types. On the south side is Pelona schist, recognizable by its large quantities of mica. On the north are granitic rocks. Within the actual fault zone may be seen several exposures of reddish sedimentary rocks that are entirely fault-bound and form elongated slices.

DEVIL CANYON

You can drive up to the picnic grounds on the San Andreas Fault at the entrance of Devil Canyon, but the

R. C. Frampton

North of San Bernardino, the San Andreas Fault (center) separates the San Bernardino Mountains from the plain to the south. The photo was taken in the early 1960s, and shows orchards as they existed before encroachment of residential development. Some of the newer developments have taken the fault location into consideration when setting street patterns and lot locations. The Mill Creek Fault (right) follows a westerly course and joins the San Andreas at Waterman Canyon. To the south, the San Jacinto Fault (upper left) veers northwest.

public road ends near there. On the hills above, the San Bernardino Tunnel and Intake Towers of the California Water Project (see page 157) are being constructed, and the area is closed to the public. The exact location of the fault in this area is well-known, since earthquake danger has played such a large part in the design and construction of the water project facilities.

The Devil Canyon Power Plant has been constructed just north of the fault, so the penstocks will not cross the fault line under high pressure. (For general information on precautions taken during construction of the California Water Project, see page 157.)

To reach Devil Canyon, take the Interstate 15 off ramp to the California State College campus, and then stay to your left on the paved road that circles the main buildings. The line of most recent activity within the fault zone is marked by a narrow trench, escarpment, and a line of brush and trees along the base of the main range.

CABLE CANYON

Cable Canyon is well known because of its generous offset along the fault line. You can see this offset clearly from the freeway—the canyon cuts a straight line up the ridge for some distance then suddenly appears to come to a dead end. The continuation of the canyon is several hundred feet to the east, and the two sections have been offset by long-term intermittent movement along the fault. West of Cable Canyon, several other smaller ravines have been similarly offset by the fault as it follows the base of the range.

TOKAY HILL

On a clear day you can sight for some distance along the San Andreas Fault from the end of Knoll Street on Tokay Hill. Plan on being there in the late afternoon when the light is low. The long shadows will accentuate the fault line and more subtle features.

SAN JACINTO FAULT

The San Jacinto Fault contributed substantially to the founding of the city of San Bernardino. Movements along the fault zone have formed a huge underground dam (known locally as the Bunker Hill dike) that has trapped great quantities of water, and the reservoir that built up behind this dam was tapped for a water supply by the city's founders. San Bernardino's well

Clarence R. Allen

Near Forest Home, Mill Creek has carved a straight canyon along the route of the Mill Creek Fault. You can drive east along the canyon as far as Fallsvale.

pattern indicates that much of its water supply comes from the area northeast of the San Jacinto fault line.

Surface evidence of the fault can be found on Colton Avenue, about a mile north of the freeway. Just north of the Coburn Avenue intersection, a low double-sided scarp rises abruptly from the flat terrain (a church is situated on top). On the north side of Colton Avenue, old DeSienna Hot Springs is located along the scarp line, and several buildings on the San Bernardino Valley College campus are thought to lie right on the active trace of the fault.

GLEN HELEN FAULT

The Glen Helen Fault is sub-parallel to the San Andreas system and lies at the base of Lower Lytle Creek Ridge. It forms the southern edge of Cajon Wash.

A very obvious scarp of this fault crosses Devore Road near the headquarters buildings of Glen Helen Regional County Park, and continues southwest through the park itself.

One of the nicest examples of a sag pond in southern California is at the base of the scarp at the beginning of the "Swamp Ecology Trail." (The larger lakes in the park are man-made.)

West of the freeway through Cajon Canyon, a small stream has been substantially offset by recent movements along the San Andreas Fault. Lost Lake, at right, is a sag pond created by the same movements.

Cajon Canyon to Palmdale

THE NOW-UNIFIED SAN ANDREAS FAULT CUTS THROUGH SPECTACULAR SCENERY IN THIS HIGH COUNTRY, AND PROVIDES AUTO EXPLORERS WITH UNEXCELLED VIEWS OF THE FAULT.

NORTHWEST OF DEVORE, the new Interstate 15 freeway cuts a wide swath through the San Andreas Fault, and the crushed and distorted rocks that typify the fault zone are evident in the road cuts. But the best views of the fault in this area are still available along the old highway route and in the hills to the west.

Road cuts on the old route clearly show the juxtaposition of rock types. North of the fault are granitic rocks overlain with marine sediments. Within the fault zone (here almost a mile wide) are exposures of Pelona schist in a distorted form, crushed and discolored by the physical and chemical changes brought about by fault movements. These crushed rocks can be examined at close range at Blue Cut near the former truck weighing station. South of the fault is an exposure of Pelona schist in its more natural state, with a regular pattern of layers similar to the sedimentary rocks.

You can best view these differences in rock types from one of the side roads on the west side of the old freeway (turn off on Cajon Campground Road.

On this west side, fault activity is also reflected in the course of a young stream that has been offset horizontally along the fault (see photograph). The stream curve is hidden by trees and brush, but by taking a short hike along its usually dry bed you can see the degree and distance of the offset.

LONE PINE CANYON

West of Cajon Canyon, the fault zone is marked by a clearly defined series of features up Lone Pine Canyon. Less than half a mile from the freeway is Lost Lake, a sag pond caused by the complex faulting that has cut off all possible drainage routes and created a bowl that serves as a reservoir for the surrounding slopes. There are visible scarps on both the north and south sides, and the pond is part of a sunken block between the two lines of disruption.

The road up Lone Pine Canyon is unpaved but maintained in good condition. It follows the fault valley and affords the motorist a rare opportunity to drive within the fault zone for some distance. An alternate route is available by returning to U.S. Highway 66 and continuing north 3 miles, then heading west on State Highway 138 for 1 mile to Swarthout Road, which leads back to the fault zone along a paved road.

Half a mile west of Lost Lake, the most recent line of faulting shows up as a low scarp and shallow trench on the south side of the road. This is evidence of very recent activity, and probably represents the rupture of 1857. There is no doubt that the scarps and troughs are fault-caused, since they are in direct contrast to all drainage features.

A little over a mile west of Lost Lake, pink sedimentary rocks begin to show up as outcroppings on the north side of the canyon. These conglomerate sandstones are known as the Cajon Beds, and they form half of a remarkable offset along the San Andreas Fault. At this point the pink rocks are on the north side of the fault. At the Devil's Punchbowl, 25 miles west, what is thought to be the same rock formation is on the south side of the fault. The two matching exposures have been separated by movement along the fault.

The few ranches located in the higher reaches of Lone Pine Canyon depend on water from springs that have been caused by faulting. Less than a mile northwest of the Swarthout Canyon Road intersection, recent scarps can be seen on the south side of the canyon behind the Sharpless Ranch, and 2 miles

Barrel Springs Road lies within fault valley.

Complex fault activity that has taken place along the San Andreas system is shown in road cuts along Cheseboro Road.

Little Rock Creek has been offset 1.5 miles by horizontal fault movement.

Tilted and weathered rocks of Devil's Punchbowl are thought to be a displaced match of the Cajon Beds formation (see below). You can explore the area on good hiking trails.

The San Jacinto Fault is well-exposed at Vincent Gap. Road between the Gap and Devil's Punchbowl is steep but colorful.

The San Gabriel River branches into Prairie Fork and Vincent Gulch branches, both of which flow along the San Jacinto Fault.

Highest elevation along the San Andreas Fault is 6,862 feet at Big Pines Summit.

The Fault is traceable through Lone Pine Canyon by a series of low scarps, sag areas and isolated springs.

It is believed that the Cajon Beds (photo on page 75) are part of the same rock formation that forms Devil's Punchbowl, and that the two sections have been separated by horizontal fault movements.

farther west, the Clyde Ranch is located on a spring that resulted from the damming of underground drainage of the canyon to the southwest.

As you drive Lone Pine Canyon Road, it quickly becomes noticeable that the straight lines of the canyon are quite unlike the ambling nature of most creek-carved features. The linearity is caused by the erosion along the zones of crushed rock within the fault.

In this area, the fault zone is at least as wide as the canyon itself, the most recent breaks being marked by the low scarps.

Two miles west of Clyde Ranch, the road makes a short loop to the south away from the main fault line, and the gray, mica schist of the fault's southern boundary begins to show up in road cuts. The road returns to the fault zone at the canyon summit. For a good view back along Lone Pine Canyon, pull off at the summit and walk northeast along the short ridge (which, incidentally, is believed to be an ancient scarp). In the distance is the San Bernardino Valley.

SWARTHOUT VALLEY

Just past the summit of Lone Pine Canyon, the fault trace and the main road enter Swarthout Valley, which is another fault-controlled feature. At the eastern entrance of the valley, many traces of the fault were completely covered by the mud flow of 1941.

Long-term action of the San Andreas Fault was an indirect cause of the mud flow that took place during the week of May 7, 1941. Over a long period, fracturing along the fault weakened the blocks of schist high on the rugged north face of Blue Ridge. The rapid melting of snow in 1941 caused the crushed rock to become top-heavy and break away at a 7,000-8,000 foot elevation and flow down Heath and Sheep canyons for nearly 15 miles. (Extensive flood control channels have been constructed to guard against recurrence).

West of Wrightwood, evidence of faulting can be seen south of State Highway 2, where a line of trees and a swamp area mark a large area of impounded ground water on the south side of the fault. The sag area 1.4 miles from Wrightwood marks the line of most recent faulting. Additional sag areas can be seen along the road to the northwest.

BIG PINES

The San Andreas Fault reaches its highest elevation in California at Big Pines Summit, 4 miles west of Wrightwood. The elevation at the road intersection is 6,862 feet. The saddle at Big Pines is fault-controlled, and the most recent trace is behind the ranger station.

SAN JACINTO FAULT

The San Jacinto Fault (see page 58) has created some very interesting topography in the San Gabriel Mountains south of Big Pines and motorists can sample the work of this great fault line by taking a short side trip south on Angeles Crest Highway to the lookout at Inspiration Point. As the legend at the Point indicates, the San Gabriel River is visible as it comes north and then splits into the Prairie Fork and Vincent Gulch branches. The location of these branches is due to the San Jacinto Fault, which cuts through the mountains here and invites the water to flow along its easily eroded course.

For a good look right inside the San Jacinto Fault, continue south for another 3 miles to Vincent Gap, where Angeles Crest Highway crosses the fault. A brilliantly colored line of red sedimentary rocks butted against gray granitics marks the fault line. These red rocks are quite out of place in this terrain, having been caught up in the fault movement and shifted around within the San Jacinto zone.

APPLETREE CAMPGROUND

One of the best exposures anywhere along the San Andreas Fault lies just southeast of Appletree Campground. A creek has cut a deep notch through the fault zone, exposing granite that has been crushed to powder by the fault movements.

To reach this exposure, park at Appletree Campground and follow the road behind the gate at the eastern end of the parking area (if the gate is open, you may be able to drive). Half a mile from the campground (a few hundred feet beyond the water tank) a steep south-facing escarpment of whitish rock marks the line of most recent activity along the fault. Water coming off the higher slopes has worn a deep notch through the scarp, and the granitic interior lies exposed. But this is not granite in the usual sense. If you hit it with a pick or other sharp object, it crumbles at the slightest blow. This is rock in name only—pulverized to an unrecognizable consistency by the great shearing action of the fault.

The significance of this exposure lies in the fact that it is probably representative of the rock conditions below the surface in this area.

The spectacular, weathered shapes of the Cajon Beds can be seen at close range on both sides of Highway 138. The rocks have been tilted by fault movements, then eroded into hogbacks.

Just southeast of Appletree Campground, a recent scarp probably dating from the 1857 earthquake breaks the mountain slope. A creek has cut through the scarp, exposing a spectacular area of severely crushed granite.

JACKSON LAKE AREA

West of Appletree Campground, Big Pines Highway follows along the fault valley. A mile to the west is Jackson Lake, which was probably a fault sag originally, but has now been substantially modified and improved into a recreation area. Just opposite the lake, there is a good exposure of crushed rock in the northern road cut.

Almost 2 miles west of Jackson Lake, you get an excellent view out over the Mojave Desert, with the Tehachapi Mountains and Garlock Fault in the background.

Just west of the 204th Street East intersection, Caldwell Lake forms a very large sag pond south of the roadbed.

THE DEVIL'S PUNCHBOWL

Five miles west of Caldwell Lake, Big Rock Creek Road leads south into the Devil's Punchbowl, a spectacular exposure of weathered conglomerate rocks that were visible as the Cajon Beds in Lone Pine Canyon. This time the rocks are on the south side of the fault—the two identical formations have been separated about 25 miles by movements along the San Andreas Fault.

The Devil's Punchbowl formations also abut the San Jacinto Fault, which at this point is very near the San Andreas Fault.

The Devil's Punchbowl has been developed as a Los Angeles County Park and is ideal for the casual hiker. The San Jacinto Fault is clearly exposed at the base of the cliff on which the Devil's Chair is located.

VALYERMO

Just west of the Angeles National Forest Ranger Station at Valyermo, a line of trees and springs along a scarp can be seen from the highway. (You can also catch glimpses of the rocks of the Devil's Punchbowl, which are remarkably similar to the Cajon Beds.)

The scarp appears to be fresh, and may date from 1857. Near the ranger station it faces north; a few hundred feet to the west it suddenly faces south, with its sloping back side turned toward the road.

In the area west of Valyermo, there is no single road that follows the fault line. Motorists will have to pick their way among several short dirt and gravel roads that wander among the low hills. However, all are marked and the route is not difficult.

LITTLE ROCK CREEK

To continue along the fault, return north up 106th Street to State Highway 138 and turn west. The highway crosses Little Rock Creek 1.5 miles west of Littlerock. This creek bed is a classic example of offset drainage along the fault. However, the length and degree of offset are so great that it is impossible to see at a glance. From its intersection with State Highway 138, the stream bed goes south for half a mile, makes a sudden sharp turn to the west as it enters the fault zone, then continues in this direction for 1.5 miles before turning south again. This long offset possibly represents some 45,000 years of intermittent movement.

Motorists can catch up with Little Rock Creek by turning south from State Highway 138 on Cheseboro Road, which crosses the fault just north of Barrel Springs Road. A mile south of the Barrel Springs Road intersection, watch for distinctive road cuts next to a line of trees and brush that crosses Little Rock Creek. The line of vegetation indicates a line of impounded ground water along the fault, and the road cuts tell a story of the faulting action (see illustrations). To geologists this fault is known as the Carr Canyon Fault; in reality, it is only a splinter of the San Andreas Fault system.

BARREL SPRINGS ROAD

To follow the main line of the San Andreas Fault, retrace your steps 1 mile to Barrel Springs Road. On the way, watch for fault breaks in the low hills to the east. The San Andreas and San Jacinto faults are very close together here, forming a wide and complex zone. Turn left on Barrel Springs Road, which follows the fault valley. A series of very low scarps between the road and the main ridge of hills to the north provides good evidence of recent activity. At one point, a line of cottonwoods marks the fault break. From high points along the road, motorists can sight down the straight fault valley to the west, past Palmdale reservoir.

At the intersection of Pearblossom Highway, continue straight on the dirt road that stays inside the fault trough for another 1.5 miles. Then turn left on the old Harold-Palmdale Road and follow it to the intersection with the Sierra Highway. At the stop sign, watch for a series of 2-foot white posts, parallel to the railroad tracks, between the tracks and the highway. These posts identify bench marks set up by the U.S. Coast and Geodetic Survey to measure possible movement along the fault.

THE SAN ANDREAS FAULT IN CLOSE-UP CROSS-SECTION

Cheseboro Road cut provides an unparalleled cross-section of fault activity in San Andreas zone. Vertical offset of gravel beds is result of recent movements. Older fault movements brought granites and sandstones into direct contact. Sketches below show one way in which the present rock alignment could have developed, although horizontal displacements may also have been important.

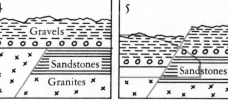

Gravels

BRANCH FAULT

Total displacement during past 1 million years

Crushed granitic rock
Approximate age: 100 million years

Silts and sandstones
Approximate age: 15 million years

1. Granites are oldest rocks. Sandstones were deposited on top.

2. Ancient faulting caused uplift of the southern block.

3. Erosion flattened surface.

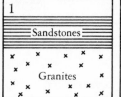
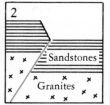
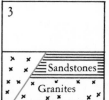
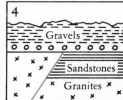

4. Recent gravel deposits covered granite and sandstone.

5. Latest fault activity uplifted northern block. Erosion rounded surface again, and red line shows section now visible in the road cut.

BRANCH FAULTS

Crushed granite

Silt and sandstones

On the opposite (east) side of Cheseboro Road, the fault zone clearly separates granites from sandstones. In Little Rock Creek (background), the impounded ground water south of the fault has caused vegetation to grow along a straight front.

In Leona Valley, recent activity along San Andreas Fault has offset a number of small stream beds north of main road.

Palmdale to Gorman

DETAILS OF FAULTING ARE SCARCE ALONG THIS STRETCH BECAUSE OF EROSION, BUT MAIN ROADS FOLLOW THE FAULT VALLEY AND TELL-TALE SIGNS ARE THERE TO SEE.

THE NEW U.S. HIGHWAY 14 (Antelope Valley freeway) crosses the San Andreas Fault just west of Palmdale. Where the highway cuts through the escarpment along the most recent trace of the fault, you find a very vivid illustration of the rock crushing and folding effects of fault action.

For a longer range view of the fault in this area, stop at the pullout along the freeway south of the Avenue S off ramp. A plaque has been erected that shows the location of the fault line and other notable geographic features. From this point, you have an excellent view of the fault valley, particularly when the light is low. The scarp marking the line of recent activity within the fault zone forms the northern edge of Lake Palmdale. This body of water probably started out as a sag pond, but it has been considerably modified as a recreation area.

To follow the fault zone west out of Palmdale, take Avenue W, which becomes Elizabeth Lake Road just outside of town.

In Amargosa Creek Canyon, the fault lies along the low mountain ridge south of the main road. Evidence of the most recent break—probably in 1857—shows up sporadically as a narrow shelf near the crest.

LEONA VALLEY

In Leona Valley, the fault assumes the form of a wide valley that is divided lengthwise by a low ridge of sedimentary rocks that have been squeezed up by the faulting action. The ridge is bounded by faults on either side; the road through the valley follows the southern branch.

Half a mile west of the town of Leona Valley, recent fault activity has formed a side hill ridge in the mountain slope. Almost 2 miles from town, where the road veers south of the main fault line, a number of offset stream beds are visible to the north.

ELIZABETH LAKE AREA

Elizabeth Lake Road continues to follow the fault west of Leona Valley. A quarter-mile north of the San Francisquito Road intersection, the road crosses the southern branch of the fault, then the center ridge, and then the northern fault branch. This, in effect, takes you through a cross-section of the fault development in this area.

The scarplets and sag areas that mark the southern branch of the fault behind Elizabeth Lake suggest that the latest movement here was within historic times, possibly in 1857. West of Elizabeth Lake, the center trough disappears from the fault zone and the two branches merge.

At the northern end of Munz Lakes, the Elizabeth Lake tunnel carrying water from the Owens Valley aqueduct to Los Angeles crosses the fault line at right angles at a depth of some 300 feet.

PINE CANYON

Past the lake area, details of faulting become scarce. Any features of recent activity that may have existed in Pine Canyon have been destroyed by erosion. The fault is still traceable geologically, but the only clear-cut visual sign is the straight fault-controlled valley that forms a natural route for the highway.

Nearly a mile west of the Pine Canyon Ranger Station, there are several good exposures of crystalline rocks that have been crushed and chemically altered by the faulting until they have become little more than white powder held together in loose clumps.

OAKDALE CANYON ROAD

West of Three Points, Oakdale Canyon Road follows the fault zone for about six miles. Half a mile west of Horse Trail Campground, a small enclosed depression and a scarp facing the mountains are visible just north of the road.

STATE HIGHWAY 138

West of the Old Ridge Route intersection, the most recent trace of the fault parallels State Highway 138 (Lancaster Road). Across the road from Quail Lake —which has been modified as a reservoir for the California Water Project—the fault trace lies right at the base of the hills.

A number of small sags and discontinuous scarps become visible north of the highway near Gorman. The line of ponds is most noticeable in the last two miles before State Highway 138 merges with State 99.

The fault passes through Gorman and then heads toward Tejon Pass. A line of telephone poles follows the fault line through a saddle cut in the slopes above the freeway.

Several dark outcroppings of volcanic rock are visible in the hillsides above Gorman. These rocks have been cut away from their original source somewhere below the earth's surface and moved around by the faulting and now appear as isolated bits of foreign material scattered along this part of the fault.

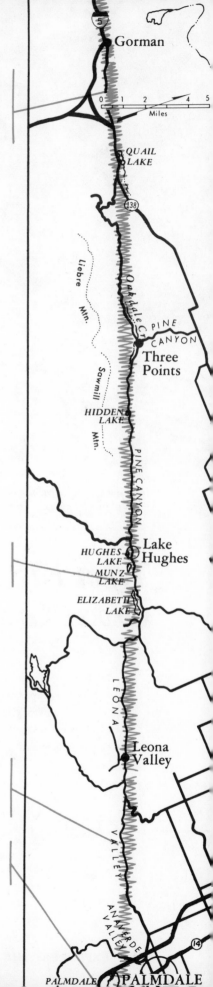

Between Quail Lake and Gorman, there are several conspicuous sag ponds and a discontinuous scarp line that point out the most recent trace of fault activity.

Near Munz Lakes, a tunnel carrying water from Owens Valley aqueduct crosses the fault at right angles.

In Leona Valley, the fault is made up of two parallel branches, separated by a ridge that has been squeezed up by fault activity.

Scarp along a recent line of activity forms the northern edge of Lake Palmdale.

Interstate 5 cuts right through the San Andreas Fault at Tejon Pass. The line of most recent activity shows up as a band of black "gouge" visible in the road cut.

Tejon Pass to Grocer Grade

IT IS IN THIS SECTION THAT THE FAULT MAKES ITS GREAT BEND TO THE NORTHWEST AND ASSUMES THE COURSE IT WILL PURSUE FOR THE NEXT 350 MILES.

IF YOU PLAN TO EXPLORE the San Andreas Fault west of Tejon Pass, it's a good idea to plan a complete route ahead of time. The road through San Emigdio Canyon comes out near Cerro Noroeste Winter Recreation Area. The nearest town is Maricopa, 30 miles to the north. The next extension of the fault is in Carrizo Plain, a lonely and desolate spot. So check the route and the gasoline supply before you start.

TEJON PASS

Until the new Interstate 5 freeway was constructed across Tejon Pass, very little of the San Andreas Fault could be seen at the summit. But the deep road cuts on the new freeway have carved a slice right out of the heart of the fault zone, so that the subsurface characteristics are clearly exposed to motorists. The fault zone generally is characterized by crushed and distorted rocks, and a band of black "gouge" clearly defines the line of most recent activity.

For the best view of the Tejon Pass exposure, leave the freeway at Gorman or Frazier Park turnoffs and drive to the summit on the old highway. There you can park and look down right into the fault zone.

CUDDY CANYON

Because of the variable widths and sometimes unrecognizable character of the San Andreas and Garlock Faults, it is impossible to pinpoint any spot where they come together. However, a reasonable estimate of the meeting ground is the area south of Frazier Park Road about three miles west of the freeway. There is no visible sign of the junction, but beneath the surface gravels, all of the rocks have been chewed to rubble by the grinding action of the left-lateral Garlock and the right-lateral San Andreas movements.

West of Lake of the Woods, Cuddy Canyon meets two valleys. Cuddy Valley extends to the northwest and is controlled by the San Andreas Fault; the valley leading southwest is controlled by the Big Pine Fault. If the Big Pine actually is an extension of the Garlock Fault, then the two sections have been offset about six miles.

There is little visible evidence of the Big Pine Fault, except for an exposure of contrasting red and white rocks visible high on the slopes north of Chuchupate Ranger Station. A branch of the Big Pine system is the dividing line between the different rock formations.

If you plan to follow the San Andreas Fault through San Emigdio Canyon, inquire at the Ranger Station about road conditions, particularly if there has been recent rain.

CUDDY VALLEY

You can see several conspicuous fault features along the road through Cuddy Valley. The first of these is a very low scarp visible at the base of the straight ridge of mountains on the south side of the valley. There is also a series of sag ponds that mark the line of most recent activity along the valley floor. An older line of rupture is indicated by the straight line of trees at the mountain edge.

At the western end of Cuddy Valley, just before the road enters the pines, a well-defined scarp is visible on the south side of the road. Its steep contours indicate recent movement, and at least part of the up-thrust on the north side was probably caused by the

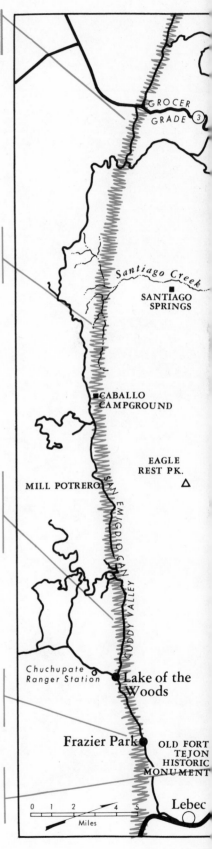

Fault scarps here are low and rounded, but stand out clearly in cultivated fields.

The fault passes through Santiago Creek Canyon; several recent features are visible from the road (photo on page 83).

Within Cuddy Valley, several scarps and sag ponds are visible from the road. One fresh escarpment starts near the west end of the valley and continues through San Emigdio Canyon.

The San Andreas and Garlock faults intersect near Frazier Park, but no details mark the exact meeting ground.

Fort Tejon was the hardest-hit area during the great 1857 earthquake in Southern California. Almost every building was knocked down.

SAN ANDREAS FAULT →

Cuddy Valley

Big Pine Fault Valley

R. C. Frampton

1857 earthquake. The valley road crosses the scarp just west of the intersection with Camp Bethany Pines Road.

SAN EMIGDIO CANYON

Before it became popular, San Emigdio Canyon was one of most isolated and pleasant stretches along the San Andreas Fault. But during the early 1970's, this area is due for extensive development. The new road through the canyon is located for the most part on a bench within the fault zone, but all details are lost under the new construction.

CABALLO CAMPGROUND

Caballo Campground is located within the fault zone, and a steep escarpment is visible just south of the main camping area. This, again, is probably an extension of the Cuddy Valley and San Emigdio Canyon escarpments.

SANTIAGO CREEK CANYON

Two miles northwest of Caballo Campground Road, there is a pullout off the main highway that affords an excellent view into the fault zone in Santiago Creek canyon. The mountain walls show many traces of fault activity (see page 83).

Aerial photograph shows the relationship of the Big Pine and San Andreas fault valleys. Big Pine apparently is an offset extension of the Garlock Fault, which intersects the San Andreas zone six miles to the east.

Jumbled topography, indicating ancient fault lines

Landslide scar

Gray granitic rocks north of fault

Most recent trace of faulting

Bright red conglomerates south of fault

View into Santiago Creek Canyon from Cerro Noroeste Road shows several traces of fault activity. Most recent break (1857) cut a narrow bench along north wall of canyon. Higher up on the slopes, a huge landslide scar and jumbled rock types are evidence of older fault movements. Repeated shifts have exposed bright red conglomerates north of the canyon (foreground) in direct contrast with the gray granitic rocks on the other side of the fault.

Exploring the Central San Andreas Fault

BETWEEN THE STARK, FLAT CARRIZO PLAIN AND THE FERTILE VALLEYS AND GREEN MOUNTAINS NEAR HOLLISTER, THE FAULT PLOWS A STRAIGHT AND NARROW COURSE.

IN THE CENTRAL SECTION, the San Andreas Fault pursues a straight northwesterly course from Carrizo Plain to Hollister. Carrizo Plain is the best place to see detailed features of the San Andreas Fault; its arid climate and lack of commercial development have helped in preserving the fault trace in almost undisturbed form.

To the northwest, the fault coincides with the valley of Cholame Creek for some distance, then passes through the higher slopes of Mustang Ridge. Between Peachtree Valley and Hollister, the fault lies east of the Gabilan Range and follows the canyon of San Benito River for a short distance.

In general, the San Andreas Fault is straight but not always definable in width. Parts of its course are obscured by landsliding, and the area south of Hollister is further complicated by the branching away of the Hayward Fault.

Even though both the 1857 and 1906 earthquakes were felt in this Central Section, no great quakes have been centered here during recorded history. The area is, however, well known for the frequency of light and medium shocks. The towns of Parkfield and Hollister, in particular, have had far more than their share of shaking during the past 150 years. Most recently, Parkfield was the scene of a significant quake in 1966.

In addition, two important areas of slow creep along the San Andreas Fault are found in this section. The winery south of Hollister is discussed on page 21; a similar creep has been recorded in the Cholame area near State Highway 46. These two instances of slow movement may be significantly related. Even more important, this steady adjustment in the earth's crust, together with the abundance of small earthquakes, may have a great deal to do with the absence of major quakes in this section during historical times. The data on these phenomena are not yet sufficient to warrant any firm conclusions, but geologists and seismologists are keeping close watch on movements of all kinds in an effort to establish a meaningful pattern.

PRINCIPAL EARTHQUAKES

Earthquake of April 11, 1885—San Luis Obispo Maximum intensities: VIII-IX.

This earthquake is usually attributed to the San Andreas Fault, even though there is some speculation that the epicenter was actually located on the Nacimiento Fault, which parallels the coast west of the San Andreas zone. The most severe damage was done in the Las Tablas district near San Luis Obispo, where chimneys were thrown down and "it seemed as though houses would be hurled from their foundations." Reports from Visalia called this the most severe shock since 1872, and some damage was done to adobe buildings in Monterey. The earthquake was felt as far away as Marysville and Ventura.

Earthquake of March 2, 1901—Stone Canyon
Maximum intensity: IX.

Centered in Stone Canyon northwest of Parkfield, this large earthquake caused a number of surface cracks 6 to 12 inches wide and hundreds of feet long. There was no evidence of actual fault movement, but some of the cracks showed 1-foot vertical displacement. Damage to man-made structures consisted of fallen chimneys in the sparsely populated areas around Parkfield and Stone Canyon.

Earthquake of March 10, 1922—Cholame Valley
Magnitude: 6¼. Maximum intensities: VIII-IX.

The shock was centered in the thinly populated area of Cholame Valley, but it damaged several houses at Parkfield and threw down chimneys throughout the area. Within the valley, cracks 6 to 12 inches wide and a quarter-mile long were opened in soft ground. There was heavy shaking at Shandon, Hollister, and San Luis Obispo.

Earthquake of June 6, 1934—Parkfield
Magnitude: 6. Maximum intensity: VIII.

There were two distinct shocks, 18 minutes apart. The center of greatest damage was at Parkfield, where a concrete block house was demolished and brick structures were badly damaged. Chimneys fell at Stone Canyon, and miners working 600 feet underground reported they could feel the shaking.

On Middle Mountain, north of Parkfield, two zones of surface cracking developed along the crest of the ridge. The largest single crack was 9 inches wide, 55 feet long, and 18 inches deep. However, no evidence of vertical or horizontal displacement was reported.

Map shows area covered in this chapter. Page numbers within rectangles refer to maps for subchapters where area inside the boxes is described.

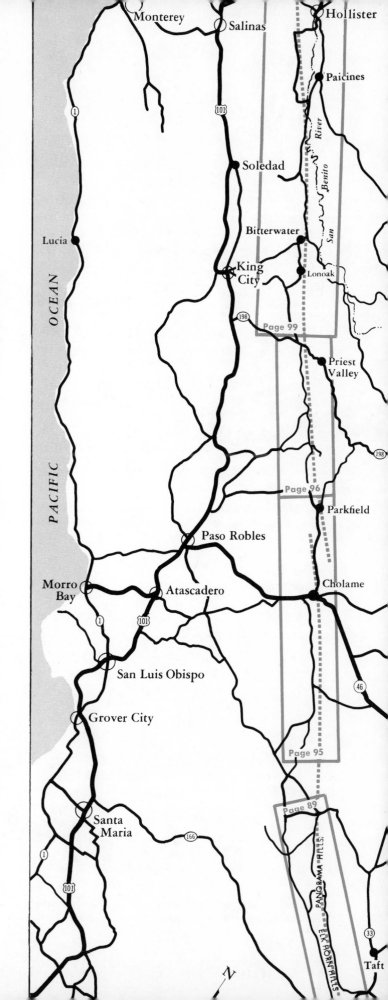

The 1906 earthquake did a considerable amount of damage in Hollister and the surrounding areas. One woman was killed when the two-story Naderman Bakery building collapsed.

Borovich Photos

Another fatality occurred when the Hollister Rochdale Company building was shattered by the earthquake force. Southern tip of the 1906 rupture was at San Juan Bautista, eight miles to the east.

U.S. Coast & Geodetic Survey

White line on Highway 46 was offset by the 1966 Parkfield earthquake. Offset was only two inches during the first day (June 22), but had lengthened to five inches (photo at right above) by August 4.

Earthquake of April 8, 1961—Hollister
Magnitude: 5.6. Maximum intensity: VII.

Many old-time residents reported this as the largest earthquake since 1906. But there were no deaths or injuries, and building damage did not exceed $250,000, even though more than half the buildings in Hollister sustained some minor damage.

The most conspicuous evidence of faulting occurred at the winery south of Hollister (see page 21).

Earthquake of June 27, 1966—Parkfield
Magnitude: 5.5. Maximum intensity: VIII—IX.

This major quake broke the surface along the San Andreas Fault for more than 23 miles. The shock was characterized by extremely high acceleration and Modified Mercalli intensities that were higher than normally expected from a 5.5 magnitude. Damage was held to a minimum only because the Parkfield-Cholame area is sparsely populated.

The main earthquake was preceded by three fore-shocks, one of which had a magnitude of 5.1. After-shocks continued for about two months. Slippage along the fault line hasn't stopped yet, ranging from about two inches the first day after the earthquake down to 1/100th of an inch per day or less after a month.

The 1961 earthquake was the largest to hit the Hollister area in many years, but damage was not spectacular. A two-story parapet was knocked over and toppled through the roof next door.

Elkhorn Scarp, a steep cliff formed by repeated fault movements, looms up at the eastern edge of Carrizo Plain. In the background is the Temblor Range.

Carrizo Plain

CARRIZO PLAIN MAY LOOK FORBIDDING, BUT IT PRESENTS THE BEST GENERAL VIEW OF FAULT FEATURES THAT CAN BE FOUND ALONG THE SAN ANDREAS SYSTEM.

CARRIZO PLAIN, THE ARID BASIN that lies between east-west State Highways 58 and 166, is the most spectacular area along the San Andreas Fault. It is also the most desolate. Only a few scattered farm buildings dot the hot flatlands. Roads are gravel or dirt and are partially covered with tumbleweeds. Rainfall is scarce, and water is seldom seen. Service stations and stores are practically nonexistent.

But it is this very desolation that makes Carrizo Plain an ideal place to study the fault. The fault movements have transformed the terrain into real "earthquake country." And because there has been a minimum of modification to the land by man and nature, these earthquake traces have remained fresh and clear through the centuries.

Carrizo Plain is the only basin of its kind in the Coast Ranges. It is a land-locked, undrained depression, 6 miles wide and 50 miles long, bounded on the southwest by the Caliente Range and on the northeast by the Temblor Range. In this basin you can see ideal examples of many fault features. There are high, old scarps that are the result of hundreds of movements, plus fresher scarplets that almost certainly were caused by the earthquake of 1857 (see page 52). The land is pock-marked by large sag ponds. And nowhere else in the world is there a greater concentration of offset drainage. For 50 miles, almost every stream channel is offset. Particularly in the southern section, the dry gullies can easily be followed as they come off the top of the hills, make sudden turns to the north just as they reach the fault, and then continue out into the plains.

To fully savor the exciting sculpture of Carrizo Plain, it is almost essential that you make your visit either early in the morning or late in the afternoon. With low light and long shadows, many of the escarpments leap up from the flat plain, and long stretches of the fault line stand out like a dark river through the rounded hills (see photo on page 43).

This is not an area for fast exploration. Many of the roads are dusty, unmarked ribbons wandering across the arid land. Landmarks are hard to locate, so don't be surprised if you make a few wrong turns and spend some extra time picking out the right fork at an unmarked intersection.

If you plan any explorations off the "main roads," be sure to carry a shovel, extra water, and plenty of gas. The jeep tracks you follow may be circuitous and unpredictable, and washouts are not uncommon.

In recent years, some of the land has been posted with "No Trespassing" signs. All of the roads mentioned here are open to the public. Do not try to explore private land without permission. Much of the Elkhorn Plain is open range, so watch for cattle and drive carefully around their watering troughs.

To help exploration, key mileages are given from the best available landmarks.

SOUTHERN SECTION

Soda Lake Road turns off State Highway 166 at Camp Dix. Before turning off into Carrizo Plain, you can get a good preview of the landscape by stopping on the highway just north of the intersection (about 100 feet north of the "Soda Lake" road sign). At this point, the road crosses a very recent scarp, and its general contour can be seen just off the roadway (especially when shadows are long). This escarpment undoubtedly was affected by the 1857 earthquake, and most of the visible elevation may be due directly to that major shake.

At the start, Soda Lake Road gives the appearance of being a paved, solid thoroughfare. But this soon changes, and you can expect a mixture of rough paving, dirt, and gravel for most of its length. Although the main fault line strikes straight across the plain along the Elkhorn and Panorama Hills, the road wanders back and forth across the lower elevations, giving the motorist several good long-distance views

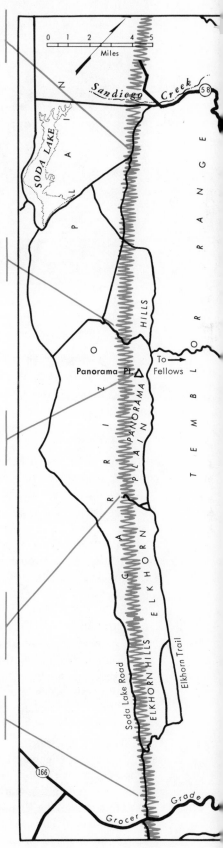

The fault is traceable across flat grasslands by series of broken scarps, offset drainage lines.

Unique double scarp formation has created a shallow valley just below the main Elkhorn Scarp.

Panorama Point provides good observation post for a general survey of Carrizo Plain. All nearby stream beds are bent along the fault line.

Where the road crosses the fault, there are classic examples of offset drainage, a fault trough, and hill slopes that have been truncated by horizontal movements.

Typical scarps and sag areas are evident near the intersection of Highway 166 and Soda Lake Road.

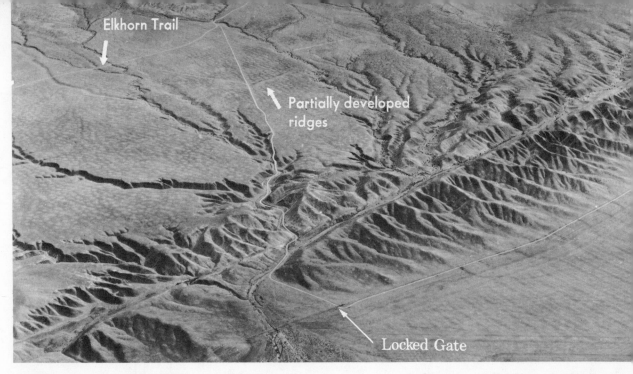

Elkhorn Trail

Partially developed ridges

Locked Gate

First stop on the route described in text is at the point where this road crosses the large river gully. The fault is easily traceable by the pronounced trough and offset streams along its course.

as well as close-ups of fault features.

At the extreme southern end of the plain, scarps and sag areas are quite evident along the east side of Soda Lake Road. One of the larger dry ponds lies at the base of a large eroded escarpment about a mile from the turn-off.

About two miles farther north (at the cattle guard), there is a good view of a high scarp to the north, and another major sag pond next to the road.

North of the cattle guard, the line of most recent activity shows up as a very low escarpment just behind the fence. When you realize that this relatively small feature was probably caused by a single great earthquake, you can understand the violent and repeated action needed to develop the huge escarpments of the Elkhorn Hills that loom just in the background.

It is most apparent here that these are very unusual mountains, jutting sharply from the plain rather than rising gradually along normal slopes. Erosion has cut many gullies into the steep face, but it is still clear that only violent upthrust could cause such topography.

About .4 mile north of the cattle guard (and 3.2 miles north of Highway 166), a dirt road turns east, crosses over the Elkhorn Hills, and intersects with the Elkhorn Trail, an unmarked but well-traveled dirt road that crosses the Elkhorn Plain in a general north-south direction. The road across the hills is steep and rutted, but suitable for most family cars. On the east

side of the hills, turn north at the first opportunity. You have a choice of routes—one stays up on the flank of the hills and leads past a water tank and cattle pen, another takes a more direct line toward Elkhorn Trail. Whichever path you take, you should wind up traveling north on the Elkhorn Trail.

About 7.2 miles north of the cattle guard near the water tank and cattle pen, watch for an unmarked side road that turns west from Elkhorn Trail and approaches the hills (it is the only side road in the area that is not posted with a "No Trespassing" sign). Follow this narrow, gully-washed road for 1.3 miles to a point where it drops into a large stream bed that is right on top of the fault (see photo above).

You can park here, walk along the fault, and take a close look at some major fault features. For a good vantage point, hike up the steep hill that forms the eastern edge of the large gully south of the road. The most recent line of faulting is defined as the narrow trough to the east of the main stream bed (and just east of your high vantage point). Looking north, you'll see that the recent line of faulting cuts a bench through the hillside just past the road. To the south, a fault trough defines a similarly straight path.

There are two good examples of offset drainage at this point:

1. The large stream gully that apparently runs north here actually is the offset section of a very old stream bed that runs west out of the hills into Carrizo Plain. The total offset is thousands of feet.

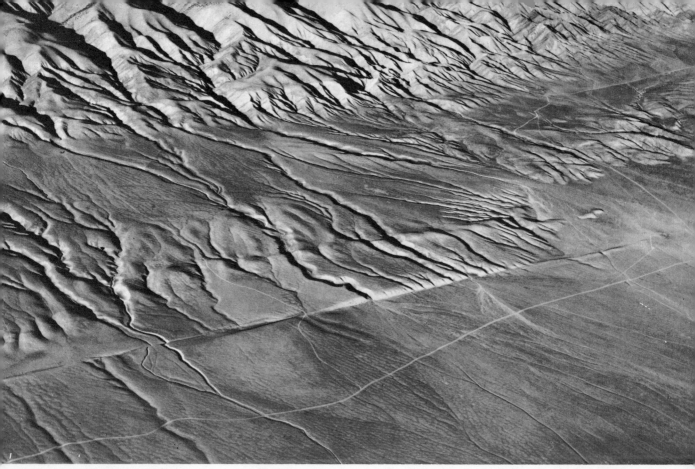

John S. Shelton

Fault can be easily followed in the northern section of Carrizo Plain. Offset drainage lines are very common (note the very pronounced jog in stream at left foreground) and scarps are visible from ground level.

Typical Carrizo Plain fault features include low, steep scarps and shallow sag ponds such as those shown here. These scarps appear so recent that they probably owe at least part of their height to movement during the 1857 earthquake.

2. The smaller stream bed that is followed by the course of the road has a smaller offset of only a few hundred feet, but the angle of displacement is clear.

If you walk south along the fault line, you will find that all stream gullies veer sharply as they hit the fault. One young gully south of the road is offset only 45 feet, but even this comparatively small offset is probably the result of several earthquakes over a 250-300 year period.

There is no great difference in rock types along this section of the fault, because the rocks are too young to show any of the major displacements found in the deposits to the north and south. However, the gravel and clay beds on the southwest side of the fault are tilted toward the plain.

From this point, retrace your route back to Elkhorn Trail. At the Y, the hiker can see more evidence of recent faulting—particularly if the sun is low in the west. The flat land here is marked by several low swells, facing east, similar to ocean waves about to break. These are particularly noticeable just south of the road. But don't look for anything dramatic—just small hillocks that rise above the regular landscape. These are, in effect, partly-developed ridges that may be thrust much higher by the next earthquake in this area.

The Elkhorn Plain is strikingly similar in character to Carrizo Plain, only it is 100-200 feet higher with the Elkhorn Scarp forming the dividing line. Most geologists feel that these two plains once were a single, broad, flat valley. But the movement along the fault has lifted the eastern section along the scarp line, thus creating two separate elevations and sets of topographic features.

NORTHERN SECTION

Seven miles north on Elkhorn Trail, Panorama Point looms up from the hills to the west; it affords the hiker an excellent view of the surrounding countryside, and there are several offset drainage channels at the fault line near here.

At the intersection with Crocker Pass Road (to Fellows), turn left and continue .4 mile to a second intersection. Take the left (western) fork that leads back toward the Panorama Hills. Follow this road 1.5 miles to the edge of the steep scarp, where you again drop back into Carrizo Plain.

To see a rare example of double scarp formation, park at the brink of the scarp (just west of the fence)

and walk around the western side of the ridge. About a quarter mile south of the road is a unique wide trough, with the higher, older Elkhorn scarp on the east and another young scarplet on the west. Both of these escarpments show signs of recent activity, and the broad gulley between them has been depressed by the uplifting action on both sides.

As the road starts down into Carrizo Plain, it crosses this trench and then stays on top of the smaller escarpment for a short distance. Just before you reach the telephone line, you can see the trench to the north, and a sag area just east of the road.

Follow the pole line for 1.5 miles and then turn right on the road that leads past the grain tanks. Four miles to the north, you again intersect the Elkhorn Trail. Between here and State Route 58, there are several points of interest along Carrizo Plain Road; as an aid to finding them, mileages are given from the Elkhorn Trail intersection.

At 1.4 miles, turn east on a dirt road for .4 mile, where the tracks mount the fault scarp at what is called a scissor point. To the south, the scarp faces west; to the north, it faces east. This type of formation usually means the two scarp faces once were

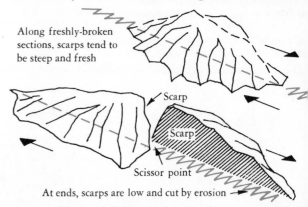

Along freshly-broken sections, scarps tend to be steep and fresh

Scarp

Scarp

Scissor point

At ends, scarps are low and cut by erosion

joined as the two halves of a low hill, but horizontal movement along the fault has shifted them into positions that resemble cliffs that appear to be caused by vertical movements.

At 2.6 miles (at the cattle guard), a branch road to the east leads along the fence for .3 mile to a classic offset stream with a displacement of 500 feet. Some 3,000 years of persistent movement probably were required to make this offset.

At 4.6 miles (at a second cattle guard), a young scarp and offset stream bed are visible to the west, indicating that there has been recent movement in this section.

Unique double scarp formation (at second stop indicated in text) indicates two lines of recent activity, as shown by red lines. Escarpment at right is much lower than that at left.

One of the classic examples of offset drainage in Carrizo Plain is this stream three miles northeast of the Elkhorn Plain-Carrizo Plain Road intersection. Offset probably represents 3,000 years of intermittent movement.

Line of recent activity in Palo Prieto Valley is marked by a low scarp that separates valley floor from low hills on th
east. Many stream channels in this area noticeably offset by horizontal movements.

Palo Prieto Pass to Peachtree Valley

MAIN ROADS DO NOT ALWAYS STAY WITH THE FAULT IN THIS SECTION; BUT A FEW SIDE TRIPS WILL BRING YOU INTO CLOSE CONTACT WITH SEVERAL NOTEWORTHY FEATURES.

THE SOUTHERN END of Palo Prieto Pass can k reached either on the partially paved country roa from Simmler or on the Bitterwater Valley Road o of Blackwells Corner. From the intersection of the two roads, the paved route through Palo Prieto Pa parallels the fault line.

The first visible sign of faulting is a very large sa pond just north of the intersection. A mile farth north, a scarp is visible behind the ranch buildings c the east. For the next few miles, more sag areas ca be seen near the road, along with a low scarp that invisible at noon but quite spectacular in low lig Features here, however, are much more rounded tha in areas farther south, primarily because of the high rainfall.

Several offset stream channels on the slope to t north are visible from the road. Watch for the gulli that bend sharply north as they cross the fault lir At a few points, the most recent line of activity marked by a bench that resembles a cow path c part way up the ridge.

Just north of Choice Valley School, the road enters the main part of Palo Prieto Pass, which has been eroded into the hills along the fault.

Still Lake and a steep scarp mark the fault line just north of Carter Grade. Though it is far from the road, the fault zone north of Still Lake can be clearly traced from the air or on maps by the straight line of lakes—Twisselman, O'Brien, Long—that mark its route.

A row of impressive scarps and offset drainage lines is visible west of Davis Road just before it meets State Highway 46 (the best example of offset drainage is 1.1 miles south of the highway). On the flat plain east of Davis Road, an oil company pipeline has been mounted on skids to provide some "give" in case a major movement takes place along the fault.

CHOLAME VALLEY

Cholame Valley is a sunken block between two active branches of the fault zone. The eastern branch is difficult to trace in the valley, but the western branch is visible as a low, discontinuous scarp at the base of the ridge bordering the valley. Where the main road to Parkfield follows this ridge, the scarp and sag areas show up at the edge of the roadway.

A scarp of the eastern fault line becomes visible east of the road as it enters the hills north of Cholame Valley. The roadway assumes a very straight course for 1.5 miles along the very edge of the steep side of the escarpment.

TURKEY FLAT ROAD

Turkey Flat Road intersects the fault about 400 feet east of Cholame Valley Road and then follows it in the canyon for a short distance. The straight northeast-facing scarp is easily seen across the creek from the road, and the stream is offset about 700 feet along the fault line.

PARKFIELD

South of Parkfield, the fault line passes almost directly under the highway bridge across Cholame Creek. The fault then parallels San Miguel Road for another mile, showing up as a rounded escarpment east of the road.

The Parkfield area has been repeatedly buffeted by many small earthquakes. The 1934 shock did its greatest damage here.

Just southwest of Parkfield, the fault apparently passes directly under Little Cholame Creek Bridge.

A low scarp and offset stream are visible where Turkey Flat Road crosses the fault.

At the northern end of Cholame Valley, the Parkfield road lies at the base of a scarp along the eastern fault branch.

The valley is a depressed block between two branches of the fault. Low scarps are visible along route of the western branch.

West of Davis Road, the fault is visible as a series of steep scarps and offset drainage channels.

A straight string of lakes marks fault's path through a range of low hills.

Palo Prieto Pass has been carved along the fault zone. Recent trace of activity lies at the base of the eastern hills.

SLACK CANYON ROAD

Northwest of Parkfield, Slack Canyon Road follows the fault zone and provides many spectacular views of the countryside. The road is paved for the first 2 miles, but then becomes a steep, graded dirt road that is impassable in wet weather and very rough during the early spring months before being graded.

The road dips through several canyons west of the fault zone and then climbs out along the crest of the Cholame Hills. Below the road are Buzzard and Nelson Creek canyons, both of which are covered with a huge landslide area along the fault zone. The fault has contributed to this landsliding by both oversteepening the slopes and shattering the rocks. The mountain slopes have been completely covered by great waves of rock and debris that have sloughed off. This is one of the most impressive fault-controlled landslide areas in the Coast Ranges. All of the normal canyon lines have been obscured by the pock-marked landslide terrain, and most of the normal vegetation has been swept off into the valleys below.

As the road climbs higher (north of Big Sandy Road intersection), there are several good views of the canyon walls along the fault line, which is generally characterized by a mixture of different rock types and colors that have been intermingled by faulting and landsliding.

The most recent trace of the fault can be seen on Bagby Ranch Road, 1.8 miles east of the Slack Canyon Road intersection. The fault shows up as two parallel valleys that are readily identifiable by their peculiar closed depressions and scarplets.

STATE HIGHWAY 198

A distinct valley marks the fault line along the crest of Mustang Ridge, east of Peachtree Valley. This valley, and two sag ponds within it, are visible from State Highway 198, 4 miles east of the intersection with State Highway 25 (Peachtree Road). Near the summit of the ridge, the road crosses the fault on fill. The sag pond next to the road has been dammed to make an artificial storage basin, while another depressed area farther south is still in its natural state.

Sag ponds and a distinct valley are visible where Highway 198 crosses the fault near the ridge summit.

East of Peachtree Valley, the fault stays high on the ridge, out of sight from the road.

Slack Canyon Road follows the fault zone; the most impressive views are of landslide topography.

The slopes above Nelson Creek (as seen from S. Canyon Road) present a classic example of landsl topography common along central section of San Andr Fault. Line shows location of scarp marking m recent trace within fault zo

Low scarp indicates location of the fault branch al the west side of Cholame Valley. At one point, main r follows on top of escarpment. Flat valley basin depressed block between two active branches of San Andreas syst

Steep scarp overlooks sag area adjacent to Coalinga Road. Most recent activity in area has been along steep eastern side. Similar escarpment visible a few miles north in Little Rabbit Valley.

Lonoak to Hollister

THIS IS ONE OF THE MOST ACTIVE AREAS OF THE SAN ANDREAS FAULT. BUT THE ZONE ITSELF IS WIDE AND DIFFICULT TO DEFINE ALONG THE FLANKING GABILAN MOUNTAINS.

NORTHWEST OF PEACHTREE VALLEY, the fault continues its straight course east of the Gabilan mountains. For several miles it follows the valley of the San Benito River.

LEWIS CREEK ROAD

Lewis Creek Road crosses the fault zone about three miles southeast of State Highway 25. The fault is characterized by landslide topography on the slope north of Lewis Creek, and by an area of severely crushed rock that shows up in road cuts south of the creek bed. Large slices of red cherts and greenstone have been turned to rubble by the faulting.

COALINGA ROAD

A few hundred yards southeast of State Highway 25 a conspicuous escarpment overlooks a narrow, closed depression adjacent to Coalinga Road. These features mark the line of most recent activity within the wide fault zone.

LITTLE RABBIT VALLEY

Two miles north of Coalinga Road, another good scarp lies east of State Highway 25 through Little Rabbit Valley. This escarpment parallels the road and faces west, so its fault-controlled steep side is clearly visible. A shallow sag area lies between the scarp and the highway. Several other low, irregular depressions and sag areas indicate the line of faulting to the north.

Less than a mile north of San Benito Road, State Highway 25 turns west out of the San Andreas Fault zone. At the sharp turn there is a good exposure of dull reddish volcanic rock in the road cut. This is normally the basement (oldest) rock east of the fault but has been raised by the fault activity until it is at the surface at this one point.

OLD AIRLINE ROAD

The Hayward Fault (see page 150) begins to split away from the San Andreas Fault near the intersection of State Highway 25 and Old Airline Road. The zone of the combined faults is 3 or 4 miles wide. An example of the complexity of the faulting in this area is the ridge that is visible just west of Old Airline Road. This ridge is "lost" in the fault zone, which is completely surrounded by the broad fault valley. The isolated ridge is made up of the same gravels that are in the young mountain ranges along the eastern edge of the fault. Presumably it has been cut away from the main ridge and uplifted within the fault zone until it now occupies a position completely out of place with the surrounding terrain.

CIENEGA ROAD

Cienega Road follows the main line of the San Andreas Fault for about 15 miles. A mile and a half north of the Old Airline Road intersection, a peculiar mountain sliver suddenly looms up from the rolling terrain east of the roadway and parallels it for about a mile. Like the island of gravels further south, this sliver apparently has been cut away from its parent mass. But here the rock belongs with the Gabilans to the west. The sliver of granite was separated from the main range by the faulting and eventually formed an isolated slice ridge.

A discussion of the Almaden Winery is given on page 21.

Hollister is frequently shaken by movements on both the nearby San Andreas Fault and the Hayward Fault (see page 150). This area also is important for the slow creep that is taking place along the fault. Evidence of the creep is noticeable on the city streets and at a winery south of town (see page 21).

Winery that is being torn apart by slow fault creep (see page 21) is located on Cienega Road.

Near Old Airline Road intersection, the combined zone of the Hayward and San Andreas faults is several miles wide.

Scarps along recent fault traces stand out clearly near Coalinga Road and in Little Rabbit Valley.

Fault movement near Lewis Creek Road has created a wide band of crushed rock, and caused many landslides.

Exploring the Northern San Andreas Fault

MUCH OF THE FAULT'S NORTHERN SECTION IS COVERED BY THICK FORESTS OR THE WORKS OF MAN. BUT SCENIC VIEWS ARE SPECTACULAR, AND YOU CAN STILL SEE MANY AREAS HIT BY THE DISASTROUS 1906 EARTHQUAKE.

THE NORTHERN SECTION of the San Andreas Fault zone extends from San Juan Bautista to Point Arena and, in effect, covers the course of the 1906 earthquake rupture.

Northwest of San Juan, the fault extends along the Santa Cruz Mountains and at times coincides with the canyons of Soquel and Los Gatos Creeks. Southwest of Monte Bello Ridge, the fault parallels Stevens Creek. On the San Francisco Peninsula, it is traceable as a wide valley that has been modified at the northern end to provide a reservoir system for the San Francisco area. One of these reservoirs is in San Andreas Valley, which gave the fault its name.

West of San Francisco, the fault lies under the Gulf of the Farallones. It re-emerges to form a dividing line between Point Reyes Peninsula and the Marin County mainland, crosses Bodega Head, and goes underwater again as far as Fort Ross, where it moves inland for the last time to parallel the coast as far as Point Arena.

The fault movement that caused the 1906 earthquake broke the surface all along this route and did extensive damage to towns and settlements on either side. Many of the details of this rupture are now gone, but the main areas of interest can still be pinpointed.

Because of the heavily timbered areas of the Santa Cruz mountains and the private property restrictions of other regions, you may have trouble getting close to the fault. But the varied topography of this Northern Section affords many spectacular long-range views across the earthquake country.

PRINCIPAL EARTHQUAKES

Earthquake of October 11-31, 1800—
  San Juan Bautista
Maximum intensities: VIII-IX.

Shocks occurred consistently for 20 days in the San Juan Bautista area. Few details are available, but historical documents indicate that as many as six a day were felt. Cracks (probably secondary) appeared in the ground near the Pajaro River.

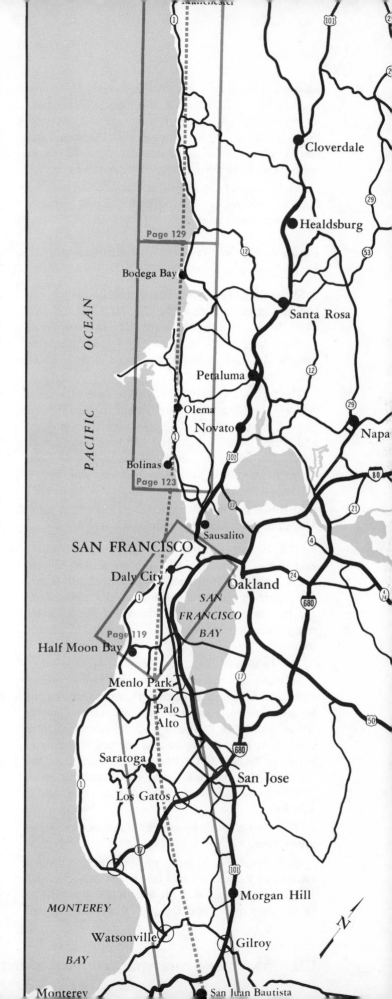

Earthquake of June 1838—San Francisco Peninsula
Maximum intensities: VIII-X.

Reports of this major earthquake, centered on the San Francisco Peninsula, are among the most confusing on record. Due to the sparcity of population at the time, there are no records that provide a clue to the exact date. In addition, details of this quake tend to become confused with another severe jolt that hit the same general area in 1836 (see page 150).

To add to the confusion, an article that appeared in a San Francisco newspaper in 1907 quoted an old-timer who had lived through a major earthquake that supposedly happened in 1839. It was assumed that this shock was unrelated to the quake of 1838, and many records include entries for both years. However, recent detailed analysis has established conclusively that the old-timer actually got his dates mixed up and that he was really talking about the 1838 quake. No tremor of any size hit the area in 1839.

Because of all this confusion, reports of actual damage have been somewhat distorted. But it is apparent that this was a great earthquake that may even have rivalled the 1906 quake in intensity along the San Francisco Peninsula. Shaking was violent at San Francisco, Santa Clara, San Jose, Mission San Jose, and Monterey. Descriptions indicate that a great fissure opened up along the San Andreas Fault.

The most vivid personal account of the earthquake comes from the same newspaper story that reported the erroneous 1839 date. For what they're worth, these are the details:

"Mr. Brown was then living in an adobe house (behind Palo Alto near Searsville Lake) . . . He had been cutting wood and had just entered the house when he was astonished by a sudden and stunning blow on the back of the head. Looking around he saw a vat which was suspended from the ceiling and which was used to hold lard swinging about the room in a most eccentric manner. Just as he was puzzling himself to account for this remarkable phenomenon, he felt the house rock and the floor tremble beneath him. Rushing to the door he beheld a spectacle of terrible sublimity. As far as his eye could reach the earth was rising and falling in solid waves . . . the redwoods rocked like to lake-side reeds. Thousands of them were broken off and hurled through the air . . .

"Mrs. Brown, at the time of the awful occurrence, was washing clothes at the side of the creek near the house. Before she was aware that the earthquake had commenced, the bed of the stream was uplifted and

Map shows area covered in this chapter. Page numbers within rectangles refer to maps for subchapters where area inside the boxes is described.

its water poured over her. Adobe houses . . . were cracked from top to bottom and fissures were made in their walls wide enough for a person to walk through. The ground was cracked in all directions and one immense opening was made . . . 10 or 12 feet in width, and its depth was never fathomed by man."

While some of the details of this account may be open to question, the general impression is certainly one of very great intensity.

Earthquake of October 8, 1865—San Francisco Maximum intensity: IX.

Shaking was heaviest in San Francisco, where two shocks caused damage to the City Hall and the area along Market Street. On the marshy lands south of Market, lamp posts and water and gas pipes were broken, and a few fissures were opened in the soft soil. There were no deaths.

The shock was fairly heavy in San Jose, where the jail and the Methodist Church were damaged. Near Santa Cruz Gap Road, chimneys were thrown down and many landslides covered the lower areas with debris. Streams near Los Gatos increased their flow, while many wells in Santa Clara County ran dry.

Earthquake of April 18, 1906—San Francisco Magnitude: 8.3. Maximum intensities: X-XI.

Not only was this one of the greatest earthquakes ever to hit California, it also was the most significant in many respects. It provided many clues to the nature of faulting and earthquakes in general, and the damage of man-made structures provided the basis for many building standards and rating tables still in use today.

The 1906 earthquake has become a popular event. It is frequently referred to as the "San Francisco earthquake," primarily because the many articles and books written about it have all concentrated on the tragic destruction of the city by the combination of earthquake and fire. But the area affected by the quake extended far beyond the limits of San Francisco, and the intensity of the shaking was greater in other areas, notably in Santa Rosa, at Stanford University, and in San Jose.

The most noteworthy aspects of the earthquake include these:

1. Catastrophe. In terms of lost lives and property damage, this quake ranks far ahead of all other California shocks combined. More than 600 persons died in the earthquake and the fires that followed, and property loss has been estimated at $400 million.

Every city between Salinas and Eureka suffered some degree of damage. The hardest hit areas included most of the major cities that lie west of the fault between San Jose and Santa Rosa. Photographs of the worst damage appear on subsequent pages.

2. Extent of Surface Rupture. The fault movement that caused the earthquake also ripped the surface for a distance of almost 200 miles from San Juan Bautista to the mouth of Alder Creek on the Mendocino County coast. This is by far the longest surface rupture ever caused by a single fault movement during recorded history.

The nature of the rupture varied, but generally it took the form of cracks and actual "rips," opened by the shearing action of the fault movement. A few low scarps were created, but the movement was predominantly horizontal.

3. Horizontal Movement. Up until 1906, geologists had generally assumed that fault movements were predominantly vertical. But this great earthquake forced a major revision in thinking, as a result of its horizontal displacements of 15 to 20 feet and scattered vertical upthrows of only 2 or 3 feet. This evidence pointed to such a drastic departure from previous thought that the official publications of the earthquake tended to hedge during the discussion of long-term horizontal movements, and it was even thought for some time that the 1906 fault movement was an exceptional event. Later measurements have indicated that horizontal shifts are the rule rather than the exception along the San Andreas Fault.

4. New Focus of Attention. Even though the San Andreas Fault had been recognized as early as 1893, its size and importance were not really understood until the 1906 earthquake. But the extent of the movements at this time awakened international geologic interest in this great rift zone. For any single fault movement to produce a horizontal movement of 20 feet was unheard of at the time, and many scientists thought that the San Andreas Fault was unique in the world. Today, it is well known that most of the master faults around the circum-Pacific basin are capable of similar horizontal shifts.

5. Contributions to Modern Seismology. After the earthquake, a State Earthquake Investigation Commission, headed by geologist Andrew Lawson of the University of California, was appointed to study the earthquake. Its report, published by the Carnegie Institution of Washington, is the greatest report of its kind ever published on an earthquake. Its pages are

still referred to by any geologist who wants to study not only the 1906 earthquake, but faulting in general. Its deep reservoir of facts has provided the research materials for countless articles and books.

This exhaustive study of the earthquake is generally considered to be one of the greatest contributions ever made to modern seismology. It set a pattern of investigation and reporting that has been followed ever since. It showed up the inadequacies of former attempts to study earthquakes, and pointed to the need for new research programs in earthquake causes and effects.

6. Elastic Rebound Theory. The elastic rebound theory, which forms the basis for the generally accepted current ideas on earthquake movement, was first stated as part of the Earthquake Commission Report. Geologist H. F. Reid examined the 1906 evidence, carefully re-evaluated triangular surveys data gathered between 1851 and 1892, and then formulated the idea of slow build-up of strain within the rocks and then the sudden release of energy when these rocks "snap" back to an unstrained position. The theory has since been modified and improved, but the basic ideas are still the same as those presented by Reid.

7. New Triangulation Networks. The U. S. Coast and Geodetic Survey triangulation networks that had gathered the first evidence of elastic build-up of stress were given great impetus by the 1906 earthquake. Surveys taken in 1906-1907 supported Reid's theory of elastic rebound, and it was suddenly realized that measurements of this type might hold the key to answering many questions about the nature of quakes.

8. Hazards of Bad Ground. The damage to man-made structures in 1906 clearly showed up the influence that bad ground can have on earthquake intensity. In San Francisco particularly (see page 38), the buildings resting on fill and "made" ground were consistently the hardest hit, while structures on top of rocky hills sustained only slight damage. This evidence was further strengthened by the damage done at Santa Rosa, which is located in a flat alluvial basin.

9. Personal Reporting. The report of the State Earthquake Commission includes hundreds of personal reports of earthquake effects and damage, which together add up to a very accurate picture of the strength and range of the shock waves.

Because of the early hour of the earthquake (5:13 A.M.), many reports from rural areas were concerned with observations made while milking cows and doing other morning chores. The amount of milk spillage out

Carl Breer

Shock waves of the 1906 earthquake caused a statue of Louis Agassiz to fall from its perch on Stanford's main quadrangle and plunge headlong through the pavement.

of shallow pans was the subject of innumerable comments and almost minute observations. This type of reporting reached its finest point at Templeton in the Salinas Valley, where it was observed that "skimmed milk was spilt at one place but unskimmed milk was not."

Since 1906, personal reports have assumed increasing importance. Today, the compilations of data made by the U. S. Coast and Geodetic Survey (see page 159) are essential to an understanding of the nature of earthquake motion.

Contrary to the usual pattern of building destruction, this row of two-story flats collapsed backward away from the street. The photo was taken before fire destroyed the entire block.

Houses along the waterfront reeled off their foundations as the filled ground beneath them slumped badly. Earthquake intensities were highest in areas of fill and "made" ground.

Within San Francisco, it was difficult to separate earthquake and fire damage. Dozens of fires broke out even while the final shudders were dying away, and the all-consuming flames quickly swept through many of the areas that had been hardest hit by the earthquake, so that sagging buildings were finished off before the extent of their damage could be recorded.

The highest MM intensity in the downtown area was probably IX or X, but there were great variations within short distances due to the differences in soil conditions (see map on page 38). It has been estimated that of the total damage done to San Francisco, only about 20 per cent was due directly to earthquake force; the remainder can be blamed on the fires.

The City Hall was one of the most notable earthquake failures. Tower was wracked and walls were thrown down by the force. Fire later spread across the area, reducing the remaining structures to rubble.

Horizontal forces generated by the earthquake caused widespread lurching in areas where the earth was poorly consolidated. Buckled car tracks, demolished sidewalks and gaping fissures in the brick streets were common.

Houses built along the water's edge at Bolinas were shaken off their pilings and dumped into the lagoon. Sand spit in the background was ruptured along the fault line.

In many respects, the real center of the 1906 earthquake was along the coast north of San Francisco. Greatest earth movements were observed in Marin County, and the city hardest hit by the shock was not San Francisco, but Santa Rosa. Virtually all of the downtown section was demolished. Investigations of the ruined buildings disclosed many instances where sound building practices had been disregarded.

Geologically, Santa Rosa is located in a flat alluvial plain—the type of ground that transmits the highest earthquake intensities. The combination of poor ground and poor construction gave the earthquake waves added impetus, even though the San Andreas Fault is 20 miles away.

Carnegie Institution

This three-story building in Fort Bragg collapsed when the sharp earthquake waves knocked out its brick walls.

Santa Rosa's City Hall and most other buildings in the downtown area were heavily damaged by the earthquake. The crew shown above is in process of clearing debris away from the collapsed main tower.

Damage was extensive along the coast as far north as Eureka. In the tiny settlement of Ferndale, all of the windows in this downtown store were broken by the shock.

Clarence R. Allen

At 5:12 a.m. on April 18, 1906, the impressive sandstone gates still guarded the entrance to Stanford University. A minute later, nothing was left but a heap of rubble.

The centers of destruction south of San Francisco were at Stanford University and San Jose. Many of the University's largest buildings and landmarks were shaken down by the quake.

Most of the other towns along San Francisco Bay were also roughed up to some degree, particularly San Mateo and Palo Alto. Agnews State Hospital was smashed to rubble, and more than 100 persons were killed.

The surface rupture along the fault extended as far as San Juan Bautista. But the area of destruction extended farther south, so that Hollister and Priest Valley were as hard hit as some of the settlements along the bay.

The arch that formed the entrance to the main quadrangle at Stanford was so damaged that the top section eventually was torn down. In the background are the ruins of Memorial Church.

Stanford's domed library took a heavy beating. The steel-supported rotunda managed to withstand the shock, but the unreinforced masonry sections were almost entirely demolished.

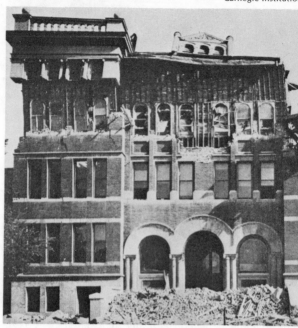

Upper walls and roof of this high school building in San Jose were knocked down by the earthquake motion.

The poorly-constructed brick main building at Agnews Hospital was thrown down by the earthquake, while water tanks in the background went unscathed.

A wood frame hotel in San Jose was knocked completely off its feet; building at left lost only its chimney.

Shock waves caused plate glass windows to part for an instant; when they snapped back into place, a corner of tablecloth was caught outside.

The 1957 earthquake punched out most of the windows in multi-story apartment building near Lake Merced but caused no structural damage.

Earthquake of March 22, 1957—Daly City
Magnitude: 5.3. Maximum intensity: VII.

This earthquake, though considerably smaller than most others reported in this book, created a great public furor since it was reputed to be the largest to hit San Francisco since 1906. While this is undoubtedly true, the quake still does not rate as very significant. Even property damage was light (about $1,000,000), considering the heavy concentration of suburban developments near the epicenter. No lives were lost. The principal damage was done in residential tracts in Daly City, where a large water reservoir was cracked. An isoseismal map appears on page 33.

Probably the worst damage done by the earthquake was this first-floor cracking of a wood frame house.

Mission

A steep scarp within fault zone makes ideal base for rodeo grandstand at Mission San Juan Bautista. In the mission garden, you can still see remains of a wall knocked down in 1906.

San Juan Bautista to Woodside

AT MISSION SAN JUAN BAUTISTA, YOU CAN STAND ON TOP OF THE FAULT; BUT IN OTHER PLACES, YOU'LL HAVE TO BE CONTENT WITH LONG-RANGE VIEWS.

A SCARP MARKING AN ANCIENT trace of the San Andreas Fault passes directly through the grounds of Mission San Juan Bautista and forms the foundation for the grandstand at the rodeo grounds northeast of the main building. The tiers of grandstand seats fit comfortably on the scarp incline, and the Mission itself is on the highest ground of the area—the top of the escarpment.

The mission was badly shaken during the 1906 earthquake and one of the walls was knocked down. Remains of this wall can still be seen just east of the rear (garden) entrance to the church. Apparently, the water-soaked ground around this wall caused it to topple while other parts of the church escaped with only minor damage.

The southern end of the 1906 rupture was located in the field north of the mission.

The area just northwest of San Juan Bautista is notable for two reasons: First, it is the southernmost locality where the ground was ruptured by the 1906 earthquake. Second, it is the northernmost locality where creep is observable along the San Andreas Fault. Fences on a ranch have been offset noticeably by the creep movement, which seems to have been going on for several decades.

This junction of creep zone and 1906 fracture zone is very significant, and lends support to the idea that the continuous creep along the middle section of the San Andreas Fault (see page 156) is gradually relieving the stress and thereby preventing major quakes such as those that hit the northern and southern sections of the fault in 1857 and 1906.

ANZAR ROAD

West of U.S. Highway 101, the most recent traces of movement in the fault zone are seen on the slope east of Anzar Road. The lines of activity are evident as a series of high terraces and side hill ridges that interrupt the normal slope.

CHITTENDEN

The fault zone generally follows close to the Pajaro River between Chittenden and the railroad bridge. The bridge was badly damaged in 1906. The concrete piers were cracked, and the track close to the river was curved in several places. At the time of the shock, a freight train was traveling south of the bridge at 30 miles an hour. About ten cars in the middle of the train were thrown off on both sides of the track. At the old railroad station, a 1,000 pound safe was thrown on its back.

In Chittenden Pass, the fault is well-exposed in the highway cuts, where it separates shale strata on the east from granite on the west. Granite in the huge Logan quarry of the Granite Rock Company has been shattered by repeated movements in the San Andreas Fault zone in this area.

CARLTON ROAD

Northwest of Chittenden, the fault trace remains high in the hills. The line of faulting can be seen from a distance, however, because of the fault-formed side hill ridges and sags that disrupt the steep face of the range. From Carlton Road, in particular, this phase of the fault topography is quite noticeable. A number of southwest-trending canyons are offset along the fault line, and there are many small lakes (not visible from the road) that are nestled in the troughs formed between the side hill ridges.

Crystal Springs Reservoir lies within a long valley of the San Andreas Fault. Highway 35 crosses the fault on a bridge separating the upper and lower sections of the lake.

Huddart Park Trail crosses the fault near the Raymundo Avenue entrance.

The Pilarcitos Fault branches away from the main San Andreas and pursues an independent course toward the coast.

You can see a noticeable change in rock types and colors where Alpine Road crosses the fault in Portola Valley.

A wide zone of alternate ridges and depressions is visible where Highway 9 crosses the fault.

At Wrights, you can still see the entrance to a railroad tunnel that was offset by 1906 earthquake. Old railroad station was on the banks of Los Gatos Creek.

From Carlton Road, you can see how fault movement has offset canyons on western slopes of the Santa Cruz Mountains.

An old fault scarp passes through Mission San Juan Bautista (photo on opposite page). Southern end of the 1906 rupture was located nearby.

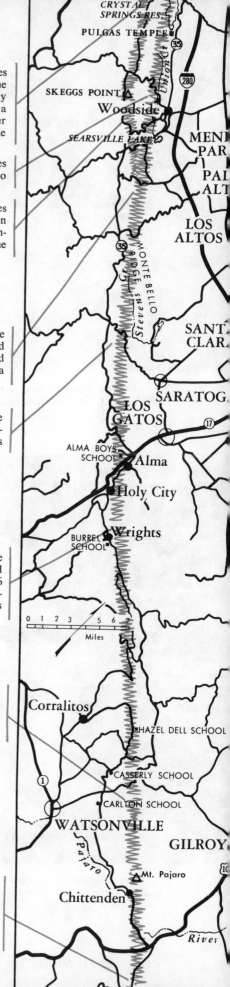

The 1906 earthquake ruptured this dirt road in front of a blacksmith shop located near the present route of Highway 17. Nearby, a house was torn in two when fault opened directly beneath it.

Intense horizontal movement of the 1906 earthquake fractured the concrete abutment of the highway bridge across the Pajaro River near Chittenden.

HAZEL DELL ROAD

West of State Highway 152, Mt. Madonna and Hazel Dell Roads come in contact with the fault zone. Mt. Madonna Road crosses one branch in a gentle swale a mile north of the Casserly Road intersection, and the main fault valley is quite visible just north of the Hazel Dell-Mt. Madonna Road intersection. A large sag area, partly occupied by Simas Lake, lies west of the road bed, and a gentle valley stretches out to the north.

HIGHLAND WAY

North of Corralitos, Eureka Canyon Road and Highland Way follow the fault zone for about six miles to Burrel Fire Station. The road is paved only part of the distance and stays high on the slopes east of the deep fault valley of Soquel Creek which cuts through the Santa Cruz Mountains. All details of recent faulting are obscured by heavy timber in this valley.

This section of the Santa Cruz mountains was the scene of many landslides during the 1906 earthquake, and the few houses situated on nearby slopes were badly damaged. One man reported that "my attention was first arrested by a slight rumbling noise; then the house trembled for 4 or 5 seconds, and this was followed by a heavy rolling motion...A heavy trembling came again for several seconds, then the heavy

Tracks of the old railroad line between Wrights and Alma were badly buckled by the 1906 earthquake force. The railroad tunnel at Wrights was offset five feet at the point where it crossed the fault.

shock that threw down the chimneys. Tables and even chairs were upset."

WRIGHTS

The fault zone follows Los Gatos Creek for about four miles. Wrights, the site of a station along the former route of the San Jose-Santa Cruz Railroad, is on the north bank of Los Gatos Creek. In 1906 this was the terminal near the eastern end of a 6,200-foot railroad tunnel through the mountains to Laurel. The earthquake ripped through the tunnel 400 feet from the Wrights entrance and caused a 5-foot offset.

The station and railway have long been abandoned. But you can still visit the site (from Summit Road, take Morrill Road east for .8 mile, then turn left on Wrights Station Road for another 1.3 miles). The tunnel entrance is a few hundred feet southwest of the San Jose Water Company's KEEP OUT sign for Los Gatos Creek. The bare face of the slope north of the tunnel is evidence of the landsliding that blocked part of the track and creek after the 1906 earthquake.

Broad fissures, cracks in roadways, and offset fences were common in this area after the 1906 shake. At the Morrell Ranch, a mile south of Wrights, a fissure opened under the ranch house. The building was thrown from its foundation and torn in two pieces. Horizontal displacements near Wrights totaled about

6 feet, and there were reports of 4-foot vertical movements.

STATE HIGHWAY 17

The fault line crosses State Highway 17 about 1.5 miles north of Summit Road. The fault valley here is characterized by more landslide scars, particularly east of the main highway.

BLACK ROAD

Northwest of State Highway 17, the fault follows a wide and deep valley occupied by Lyndon Creek. Black Road crosses to the southwest side of the fault zone about a mile from the highway turn-off, then continues to Skyline Boulevard (State Highway 35). For an excellent view of Lyndon Canyon, turn off Black Road onto Gist Road 2 miles west of Highway 17 and wind up the steep slope for a few hundred feet. The straight lines of the valley to the southeast are particularly noticeable from here.

STATE HIGHWAY 9

The fault crosses State Highway 9 four miles east of Highway 35. A number of small ridges and alternate depressions within the fault zone are visible from the road.

Old Alpine Road was badly ruptured in 1906 where it crossed the fault zone. Fracturing here was compound, and a considerable right-hand offset was created by a number of small but related dislocations.

Lake Ranch Reservoir, in the mountains south of Highway 9, near the southern tip of Sanborn Avenue, was right in the path of the 1906 trace, and the earthen dam was damaged.

To the northwest, the fault follows the canyon of Stevens Creek, southwest of Monte Bello Ridge. There are a number of pullouts along Highway 35 that afford good views across the fault valley and the surrounding countryside.

PAGE MILL ROAD

Page Mill Road crosses a pronounced fault saddle 1.5 miles northwest of Highway 35. North of the roadway is a shallow sag pond. On a clear day you can sight along the fault line all the way to Crystal Springs Reservoir. South of Page Mill Road, the fault valley along Monte Bello Ridge is clearly visible for several miles. It was near Page Mill Road that three oak trees, one with a 12-foot base, were uprooted by the 1906 earthquake.

ALPINE ROAD

You can drive along the fault zone for a short distance by turning off Page Mill Road onto the dirt extension of Alpine Road. For the most part, the 3-mile unpaved stretch stays on the southwestern edge of the fault zone.

Where Alpine Road crosses the fault at right angles in Portola Valley, there is a very noticeable change in rock types and colors. The rocks west of the fault are predominantly dark brown gravels in a clay matrix. East of the fault, the rocks are red-brown volcanics with some gray limestones showing. These volcanics were deposited some 90 million years before the gravels, but the two differing formations have been brought face-to-face by the extensive fault movement.

SEARSVILLE LAKE

The 1906 rupture cracked Portola Road at Searsville Reservoir; the general location is .1 mile north of the Family Farm Drive intersection. The belt of cracked ground was 7 to 8 feet wide.

LINEAR ACCELERATOR

Stanford University's Linear Accelerator, a two-mile-long electron "gun," built under contract with the Atomic Energy Commission, is located not far from Sand Hill Road. The western tip of the accelerator is near the intersection of Sand Hill and Woodside Roads, which is less than a mile from the San Andreas Fault zone.

Proximity of the accelerator to the fault had a great effect on engineering and design of the project. The Sand Hill Road site was selected because the accelerator housing could rest on hard bedrock, thereby less-

Small craters created by explosions of sand and water that were disturbed by violent earthquake motion were commonly found in the flat alluvial plains near Watsonville.

ening the danger from shock waves generated by an earth movement along the fault, and because it was free of any branch faults that might jeopardize the accelerator by rupturing the earth beneath it. Cuts and fills have been held to a minimum, and in some cases loose alluvial and residual soils have been stripped away and replaced with engineered, compacted fill.

Regional build-up of elastic strain in rocks adjacent to the Fault posed another problem. The copper pipe that actually forms the accelerator's electron transmission tube is 10,000 feet long and must be held in an absolutely straight line, with an allowable variation of only one-fourth inch in any 90-day period. It was originally thought that the strain build-up might interfere with this critical alignment. However, measurements indicate that the accumulated relative strain east of the fault zone is insignificant in this area, and should pose no problems over the two-mile route.

WOODSIDE ROAD

Northwest of Searsville Lake, Woodside Road parallels the main trace of the fault for some distance, then crosses it. The site of the 1906 rupture crosses the road between Tripp and Kings Mountain Road.

CAÑADA ROAD

Cañada Road lies along the east side of the fault zone, which is very wide here and includes a series of parallel ridges and valleys. These ridges may have been uplifted by the compound fracturing within the fault zone, then modified by normal erosion.

Many of the road cuts along Cañada Road, particularly near Pulgas Water Temple, expose badly sheared rocks that have been physically and chemically altered by fault movement.

HUDDART PARK TRAIL

The riding and hiking trail through Huddart Park northwest of Woodside crosses the San Andreas Fault just a few hundred feet from Raymundo Avenue. A sag area that marks one of the most recent traces within the fault zone is located at the foot bridge across West Union Creek. A shallow gully east of the creek bed suggests another branch.

To reach Raymundo Avenue from Canada Road, stay on the west side of the Interstate 280 freeway by turning left on Runnymede Road. Runnymede makes a sharp left turn about a mile north of Canada, and becomes Raymundo.

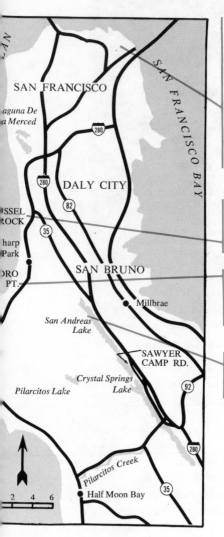

San Francisco's Ferry Building is located equidistant from the active San Andreas and Hayward Fault zones. The largest shocks that have originated from these faults — 1868 and 1906 — caused considerable damage in the city. The 1906 jolt, together with the fire that followed, stands as the greatest earthquake disaster in California history.

The fault goes out to sea at Mussel Rock, crosses the Gulf of the Farallones and emerges on land again at Bolinas Lagoon.

Pedro Point may be a section of coast that has been moved west by ancient horizontal movement along the Pilarcitos Fault.

San Andreas Fault was named after this lake, which has been expanded as part of the area's reservoir system.

Crystal Springs to Mussel Rock

MAIN ROADS FOLLOW THE FAULT VALLEY CLOSELY ON THE NORTH PENINSULA, AND ONE SIDE TRIP TAKES YOU TO THE EDGE OF SAN ANDREAS LAKE, NAMESAKE OF THE FAULT.

THE SAN ANDREAS FAULT VALLEY is easy to trace through the low hills of the North San Francisco Peninsula, since much of it has been drowned to form reservoirs for the city's water supply. Upper and Lower Crystal Springs Lakes and San Andreas Lake are all located in the canyons and deep gullies of the fault zone. North of the reservoirs, the fault is generally covered by man-made developments for 6 miles to Mussel Rock, where it goes into the Pacific.

State Highway 35 crosses Crystal Springs Lake on a low embankment that separates the upper and lower reservoirs. Prior to the completion of Lower Crystal Springs Dam in 1888, this embankment served as a water barrier between the lakes, but in 1906 it was little more than a roadway with equal water level on both sides. The earthquake caused an 8-foot offset in the embankment, but there was no loss of water.

To the north, the deep fault valley lies east of Interstate 280. The lower ridges and valleys within the main valley are completely under water. But when Gaspar de Portola camped here in 1769 (a state Historical Marker points to the exact spot), the valley was a lush area of streams and small ponds in a setting of gentle meadows and low hills. Early settlers of the Bay Area know this as a popular recreation spot, but more practical minds visualized the fault valley as the logical place to store water for a growing San Francisco. Most Coast Range valleys are too short and steep for reservoir sites; only such long straight fault-controlled valleys could be modified, extended, and dammed to make a practical reservoir system.

The Spring Valley Water Company, and, after 1930, the San Francisco Water Department, modified the fault valleys into huge catch basins, and the lakes were strung out in a straight line for 11 miles. Pilarcitos Creek, which flowed west into the ocean, was dammed in 1864 (the Pilarcitos Fault—see page 120 —now runs directly through the main body of Pilarcitos Lake and follows the creek to the south). San Andreas Creek was blocked during 1868-1870, and the dam was raised in 1875 and again in 1928. The Upper Crystal Springs Dam went up in 1877 but became less important when the Lower Crystal Springs Dam was constructed in 1887-1888. It was one of the largest concrete dams in the world at the time.

The 1906 earthquake offset the reservoir floor a thousand feet west of this huge dam but did not cause the slightest crack in the concrete—a tribute to the engineers.

This view, looking south along the valley of the San Andreas Fault on the San Francisco Peninsula, illustrates how most details of the fault line have been obliterated by residential and highway developments. Mussel Rock is in the lower right hand corner of the photo. Cliffs in foreground have been the scene of extensive landsliding.

The reservoir system traps rain water from the Peninsula hills, but it receives the great bulk of its supply from the Hetch Hetchy Water System, which includes reservoirs in the Sierra Nevada and Coast Ranges. Water from these sources flows through pipes across the southern end of San Francisco Bay and into the Crystal Springs system. The Hetch Hetchy outlet is marked by Pulgas Water Temple on Cañada Road south of the upper reservoir.

SAWYER CAMP ROAD

Sawyer Camp Road affords a very pleasant detour off the main highway and along the shores of Lower Crystal Springs and San Andreas Lakes. To reach it from the south, exit from Interstate 280 on the Half Moon Bay Road, turn north on Skyline Boulevard, and then watch for signs just north of Crystal Springs Dam.

It is difficult to define the limits of the fault zone in this area. The zone is generally about as wide as Crystal Springs Lake. The 1906 rupture ran along the extreme eastern edge of this zone.

The linear nature of the reservoirs is readily apparent from Sawyer Camp Road. Since the San Andreas Fault movement originally carved very straight valleys through the hills, the bodies of water also assume straight lines as they follow the low areas.

SAN ANDREAS LAKE

The waters of San Andreas Lake are held back by a dam on the southern end that also serves as the Sawyer Camp roadway. The dam consists of two earth embankments separated by a natural rib of rock. In 1906 this dam was somewhat lower but of the same general shape. The rupture crossed near the eastern end and fortunately cut its way through the rock rib rather than through any of the man-made sections. Rocks were shattered by the displacement and the road was offset, but no water was lost. Below the dam, on the north side, a heavy wooden flume on a trestle within 50 feet of the fault trace was demolished.

STATE HIGHWAY 35

The northern end of Sawyer Camp Road ducks under the Interstate 280 freeway and then intersects with Skyline Boulevard again. To follow the fault, turn

north, climb up onto the freeway for a short distance and then exit again on State Highway 35, which is Skyline Boulevard under another name.

Just past Sneath Lane intersection, State Highway 35 and the fault come together, offering a rare opportunity to drive "inside" the fault. The narrow gulley that forms a natural route for the road is the result of recent (several thousand years) movement within the fault. The 1906 rupture passed right through the trough. The eastern road cuts that are along this short stretch show a variety of rock types that have been crushed by the fault movement into a mass of gravels and clay.

North of Berkshire Avenue, the straight fault valley is covered by residential development east of Highway 35. The 1906 rupture tore out a large water main in this valley, thereby contributing to the desperate water shortage during the San Francisco fire.

New residential developments also hide the line of the fault from this point all the way to the coastline, except for a few sag areas near the King Road intersection.

MUSSEL ROCK

Mussel Rock is the northern limit of the San Andreas Fault on the San Francisco Peninsula. The rock lies just off the mainland and may be part of a giant landslide that conceals the fault on the coastal bluffs. The top edge of the slide area is high above the coast.

This coastal area is privately owned and sometimes is closed to vehicular traffic.

North of Mussel Rock, landslides have been so frequent in the steep slopes made up of soft sediments that the road which once followed the coast north to Thornton Beach has been permanently closed.

The epicenter of the 1957 earthquake was very near Mussel Rock, and several slides occurred on the bluffs at that time.

PILARCITOS FAULT

The coast line at Mussel Rock shows no obvious signs of long-term right lateral fault movement. However, another old fault line farther south does merge at the coast where the shore makes a large bend. The Pilarcitos Fault branches away from the main San Andreas Fault near Black Mountain west of Los Altos Hills and plows its own course northwest to the coast at Pedro Point, about five miles south of Mussel Rock.

Pedro Point juts out into the Pacific, and may represent a section of land south of the fault that has been moved horizontally into the water.

This fault line also separates two radically different rock types—the granitics that form Montara Mountain and other hills to the southwest, and the Franciscan Formation (primarily sandstones) on the northeast. The Franciscan formation lies on both sides of the part of the main San Andreas Fault lying east of the Pilarcitos.

This evidence, plus the proximity of the Pilarcitos Fault to the San Andreas, leads some geologists to believe that millions of years ago the Pilarcitos Fault was really the main line of the San Andreas. But because of the curves in the Pilarcitos along this stretch of coast, the San Andreas Fault may have broken a new, straighter route more in line with its linear character.

The Pilarcitos Fault is still active and may prove its relationship to the parent San Andreas system in some future earthquake.

The road across Crystal Springs Reservoir was offset but not destroyed during the 1906 earthquake. Fence line shows degree of displacement.

A man stands on the mole track or line of furrowed ground that characterized the 1906 rupture near Crystal Springs Lake. The fence at left was offset several feet by the horizontal movement.

CRYSTAL SPRINGS TO MUSSEL ROCK   121

Bodega Head

North of San Francisco, the valley of the San Andreas Fault divides the mainland and Pt. Reyes Peninsula on the left. In the background, the fault drops into the ocean and then crosses Bodega Head.

Bolinas Lagoon to Bodega Head

MOST OF THE COUNTRYSIDE HERE IS FENCED; BUT FORTUNATELY, HIGHWAY 1 FOLLOWS THE FAULT VALLEY WHERE THE GREATEST MOVEMENTS OF THE 1906 QUAKE TOOK PLACE.

NORTH OF THE GOLDEN GATE, the San Andreas Fault cuts a remarkably straight valley through Marin County from Bolinas Lagoon to Tomales Bay, separating Point Reyes Peninsula from the mainland.

BOLINAS LAGOON

The fault zone is wide as it enters Bolinas Lagoon. The rupture of the 1906 earthquake, however, crossed the sand spit at the mouth of the lagoon and was first traceable on the shore a quarter of a mile south of the intersection of Bolinas Road and State Highway 1. Signs on the spit were visible for only a few months, as the shifting sand dunes quickly covered the rupture. At the shoreline, the thick covering of brush completely hides any fault details.

To the north, Highway 1 follows the fault zone to the town of Point Reyes. The zone ranges from several hundred feet to more than half a mile in width, and the road generally stays within this broad belt.

HORSESHOE HILL ROAD

A narrow valley caused by recent activity (including 1906) along the fault is visible just east of the Highway 1-Horseshoe Hill Road intersection (about .4 mile north of Bolinas Road). To spot it, walk a few feet west of the intersection and watch for the narrow gulley that cuts through the trees to the northwest. (Don't be confused by the larger valley to the west, which is an ancient stream bed that has been elevated by earth movement.)

FAULT VALLEY

For the next 5 miles to the north, Highway 1 leads through an area of complex faulting. Within the fault zone is a series of low parallel ridges, separated by shallow ravines and gullies. This type of terrain is caused by complex fracturing, some "slices" of earth being uplifted and neighboring blocks depressed. The erosion that quickly starts to work rounds the higher blocks into low mounds, while the gullies are cut wider and deeper. The 1906 trace is not visible here from the road; it is on the west side of the fault zone (behind the parallel ridges), while the road hugs the eastern slopes. Some of the houses on the surrounding hills were standing in 1906. Those with foundations on solid rock sustained little damage, despite their proximity to the rupture.

The drainage of the valley has been substantially altered by the complex fault movement. At one point, two parallel streams only 1,000 feet apart drain in opposite directions. The eastern stream flows north to Tomales Bay, while the flow in the western stream is south toward Bolinas Lagoon. Each is actively eroding along an old fault trace.

POINT REYES NATIONAL SEASHORE

This is one of the best places to study the San Andreas Fault, trace the 1906 rupture, and learn about recent advances in seismology.

The main headquarters buildings of the National Seashore are located at the old Skinner Ranch, one of the hardest hit areas of the 1906 earthquake. The diagram on the opposite page shows the relationship of the buildings to the surface rupture. The three monuments were set up by the state earthquake commission right after the 1906 tremor to record any future movements (a fourth marker located east of the fault to complete a measuring "box" has been removed).

The ground moved 15 feet, 9 inches in front of the Skinner's main ranch house (see photo on the next page), so the garden path which had led to the front

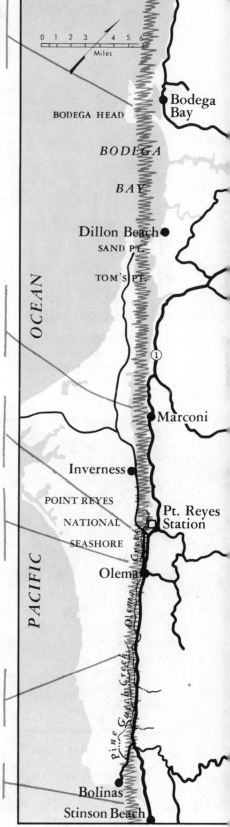

The fault separates Bodega Head from the mainland. The 1906 break passed close to the shore, but all traces were soon erased by wind and water.

The relationship of Tomales Bay to fault movements is readily apparent. The 1906 trace was under water, but caused mud flows that damaged piers at Inverness.

The greatest earth movements of the 1906 earthquake took place in this valley. The largest movement was 20 feet, across the Sir Francis Drake Highway.

Along Bear Valley Road, you can see markers that show the path of the 1906 rupture (see page 125) and a side hill ridge that resulted from fracturing during the same shock.

The drainage of this area has been so modified by fault movements that parallel streams drain in opposite directions.

Bolinas Lagoon is the direct result of erosion along the fault zone.

At the Skinner Ranch, the 1906 rupture passed beneath the cow barn at right. The larger part of the barn west of the fault held its foundation, but the smaller half to the east was dragged 16 feet.

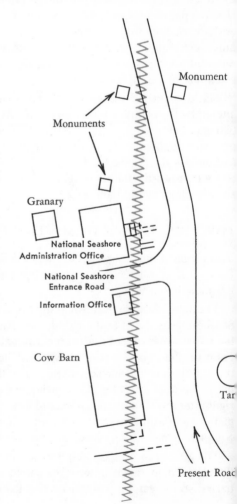

Monument

Monuments

Granary

National Seashore
Administration Office

National Seashore
Entrance Road

Information Office

Cow Barn

Tar

Present Road

Before the 1906 earthquake, the path led up to the front steps of the farmhouse. But horizontal movement of 15 feet, 9 inches moved path into position shown here. The house was not seriously damaged.

door suddenly faced a wall. However, the farmhouse itself was not permanently damaged. A dairy building (near the present Information Office) remained on its foundation, even though the rupture passed within three feet. A granary located 100 feet from the rupture was shifted off its foundation but not seriously damaged.

A cow barn that must have been located where the present red barn stands overlapped the line of rupture and was torn in two. The west side held to its foundation, but the smaller part on the east side of the break was dragged 16 feet.

The National Seashore's Information Office contains a seismograph, one of several that are now located along the San Andreas Fault zone to measure the small quakes that frequently occur. It is hoped that analysis of these "micro quakes" will provide some new insight into the nature of the big shakes.

A self-guiding Earthquake Trail has been developed on the National Seashore property to enable visitors to take a short walk along the fault zone and learn something about earthquake activity. It is the only trail of its type found along the entire San Andreas Fault. Signs along the way point out a fence that was offset 15 feet by the 1906 earthquake, a creek bed that has been altered by earth movements, and the spot where a cow supposedly was "swallowed" by the 1906 quake.

This famous event occurred on the old Shafter Ranch, just south of the Skinner property. There are two versions of the story. The first states that a cow happened to be standing along the fault at the time of the earthquake, and the poor animal fell into one of the trenches that were opened up by the twisting ground motion. When the ranchers could not remove the injured beast, they filled in the trench as a grave.

The second version claims that the earth actually opened and swallowed the cow, leaving only the tail visible above the ground.

The official report of the Earthquake Commission states: ". . . at the Shafter Ranch, a fault crevice was momentarily so wide as to admit a cow, which fell in head first and was thus entombed. The closure which immediately followed left only the tail visible."

BEAR VALLEY ROAD

To the north, Bear Valley Road parallels the 1906 trace. For the most part, the line of rupture is hidden by trees and brush, but it can be clearly seen as a side hill ridge 2 miles north of Olema. The ridge is visible on the hillside east of the road near the only speed limit sign ("Slow to 25 Miles") on Bear Valley Road.

The nature of the ridge is most evident where the water coming off higher slopes has cut a gulley into the hill since 1906 and has exposed the fault.

The 1906 movement ruptured the surface on the hillside in a mole track effect (see photo, page 127) and made a trench along the line of the break. Later erosion washed out the trench, while the lower slope held firm and eventually assumed the form of a ridge.

From this point you can grasp the width of the entire fault zone in relation to the 1906 break. The greater part of the fault zone is west of Bear Valley Road and is marked by a broad valley covered with low marshes. The 1906 earthquake cut only a minor fissure along the east side of this zone.

SIR FRANCIS DRAKE HIGHWAY

The greatest horizontal movement of the 1906 earthquake—20 feet—took place across Sir Francis Drake Highway just west of Point Reyes. The break was not a clean line, but rather a series of interconnected ruptures, each contributing to the total offset.

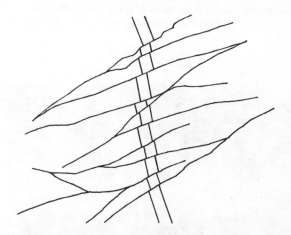

You can easily find the point of displacement, but there is little to see. The road, of course, has been straightened and repaired any number of times during the last 60 years and all signs of the 1906 break are gone. But you can stand on the exact spot just west of the main culvert across the marshes.

For a more satisfactory view, stop at the intersection of Bear Valley Road and State Highway 1 and look east along the asphalt. If you stand right in line with the road bed, and squint with the eye of faith, you can still see the great sway in the highway.

TOMALES BAY

Tomales Bay is a fault feature. Its width is slightly more than that of the fault zone, but the straight shore

Greatest horizontal movement of 1906 earthquake was 20 feet, across Sir Francis Drake Highway west of Pt. Reyes (this figure often mis-stated as 21 feet; however, original measurement was recorded as 20 feet). Photo at right shows highway today; line indicates location of the rupture (drawing on page 125).

Swath indicates the generally accepted width of the complex San Andreas fault zone at Bodega Head.

lines indicate a strong relationship to faulting. The 1906 break was under water, but it caused great mud flows that offset and telescoped several piers that jutted out from the shores at Inverness.

BODEGA HEAD

Between Tomales Bay and Fort Ross, the fault lies under the Pacific, except for its crossing at Bodega Head. Actually, the fault passes between the Head and the mainland, and is covered with the sand dunes of Doran Beach and the wide neck that forms the northern edge of Bodega Harbor.

As in many other areas, the fault separates the granitic rocks of Bodega Head and the sandstones and shales of the Franciscan formation on the mainland. Between these two formations is a wide belt of mixed and crushed rock that normally characterizes the San Andreas Fault zone. Since sand and water cover most of the fault, there are no obvious scarps or other details to mark a line of recent activity. The 1906 trace passed very close to the mainland. There was no measured displacement at that time, but movement of some kind can safely be assumed, since offsets were substantial at both Tomales Bay and the coast near Fort Ross.

*Typical evidence of horizontal movement found along
the fault line in Marin County after the 1906 quake
illustrated by this offset fence, which was broken
cleanly by the sudden movement.*

*Along Bear Valley Road, north of Olema, the 1906 trace
cut a furrow along the hillside and slightly elevated the
lower slope in relation to upper ground.*

The 1906 surface rupture passed within a mile of Fort Ross, and historic old wood church was flattened by the great force. The building later rebuilt as part of a State Historical Monument.

Fort Ross to Point Arena

HEAVY TIMBER FREQUENTLY HIDES THE FAULT ALONG THE COAST, BUT THE CURIOUS MOTORIST CAN FIND MANY ISOLATED MEMENTOS OF THE GREAT EARTHQUAKE OF 1906.

THE FAULT GOES INLAND again just south of Fort Ross and lies within 2 to 5 miles of the rugged coastline of Sonoma and Mendocino counties for 43 miles before dropping off the headlands north of Point Arena and back under the Pacific.

Through most of this section, the 1906 fault movement was noted for its multiple surface fractures. As many as six or eight parallel breaks were found in the broad fault valleys near Fort Ross, Plantation and Manchester. Horizontal displacements were on the order of 10 to 15 feet, even as far north as Alder Creek.

This was an area of great displacement and destruction during the 1906 earthquake, but the logging projects and natural regrowth of half a century have all but buried the scar in most areas. The country is rough and wooded, and the fault frequently is lost among towering stands of redwood and fir, invisible to all but the geologist who can dig into the mountain slopes and trace the lines of rock deformation. But near Fort Ross, Plantation, and the northern coastal communities, you can still see signs that identify a wide and recently active line of rupture.

FORT ROSS

The fault first emerges on the Sonoma coast at Timber Gulch, an appropriately named inlet on the high bluffs 5 miles north of Russian Gulch and 2.5 miles southeast of Fort Ross. The fault zone here is about 1,000 feet wide, and there are only general features to mark its appearance, such as an area of sag ponds west of the highway and some exposures of crushed and deformed rocks along State Highway 1 road cuts.

The roadway circles inland to cross Timber Gulch, and the fault line stays between it and the coast for a short distance before turning inland. Once into the hills, the most recent line of faulting becomes instantly recognizable as a wide and clear path of scarp formations and parallel depressions that follows a straight line through the hills.

Along the coast, Pine Gulch (.8 mile north of Timber Gulch) shows evidence of a substantial offset where it crosses the fault trace.

FORT ROSS ROAD

There are conspicuous signs of faulting on the old Fort Ross Road, leading north into the hills from the State Historical Monument. At a point .8 mile north of the Fort, a logging road branches west from Fort Ross Road; it was here that the 1906 rupture cut through the mountain slope like a giant rake, plowing as many as eight separate furrows, knocking down trees, and displacing stake fences 10 to 15 feet. The lines of rupture have partially filled with silt and debris, but they are still recognizable near the road. On the south side, there are only one or two of the gulleys, but the motion along the fault changed the course of a stream near here and thereby ended a flood threat along the base of the hill. On the northwest side of Fort Ross Road, the alternating trenches and low ridges are visible just behind the fence.

Eight-tenths of a mile north of the logging road is a giant redwood tree that had its top growth sheared off 30 feet above the ground by the force of the 1906 earthquake according to local historians. The tree managed to live, however, and new trunks now tower above the breaking point. The decayed remains of the trunk that was thrown down by the earthquake can still be seen lying at the eastern edge of the roadbed.

The fault goes out to sea at the mouth of Alder Creek. You can see the site of old bridge that was torn apart by the 1906 earthquake.

Course of the Garcia River follows the fault valley for several miles.

Both the South and North forks of the Gualala River have carved channels along weakened rocks of the fault zone.

Plantation is located in the fault valley; a tree split by the 1906 surface rupture still stands near the road.

Northeast of Fort Ross, a redwood tree sheared off 30 feet above the ground by the 1906 shock is still visible.

PLANTATION

North of Fort Ross, the San Andreas Fault stays with its straight northwesterly course, while State Highway 1 follows the coast line. The next convenient spot to see the fault is at Plantation, a quiet settlement about six miles above Fort Ross.

Plantation can be reached from Highway 1 either through Kruse Rhododendron Reserve State Park (dirt road), or by leaving the coast at Timber Cove on Timber Cove Road, turning left on Sea View Road, following the eastern rim of the fault valley for 3 miles, and then dropping down into the fault valley on Plantation Road.

The buildings at Plantation lie right in the fault zone, which assumes the appearance of a gentle valley. In 1906 there were as many as six parallel lines of rupture in this area. Two sag ponds once occupied the front yard of the houses now standing on the north side of the road. One has been drained for use as a volleyball court; the other has been filled to make a flat yard area.

Just at the eastern edge of the ranch buildings, a redwood tree that was split open by the 1906 earthquake stands on the north side of the road. The crack near the base of the trunk is still visible.

Farther north in the rift valley, two large sag ponds have been dammed and converted to recreation areas. They are fed by springs that have broken through the surface along the fault.

GUALALA RIVER

The Gualala River empties into the Pacific Ocean 10 miles north of Stewarts Point. It is the largest river along this section of coast, but more important, it is the only river in the northern Coast Ranges that parallels the coast line for any great distance. The reason behind this peculiarity is the San Andreas Fault. From a point north of Plantation to the river's mouth at Gualala, the south fork of the stream is controlled by the fault. The shattered rock within the fault zone invites water and quick erosion, and the Gualala's course has changed to fit the straight lines of the fault zone. The course of the river may be seen from any one of the side roads that leave the coast and wind into the scenic mountain country (Stewarts Point Road, Annapolis Road, or the road out of Gualala). The countryside is beautiful, but the actual slopes of the fault river valley are heavily wooded and no details are visible from the road. The Stewarts Point Road bridge over the South Fork of the Gualala was badly damaged in 1906.

GARCIA RIVER

The lower reaches of the Garcia River also follow the fault zone for 7 miles southeast of Point Arena. For a look at this valley, take Riverside Drive out of Point Arena toward the Air Force Station. A bridge crosses the river 5 miles in from the coast. The flat terrace west of the river bed is also part of the fault zone which is comparatively wide in this region.

MANCHESTER ROAD

East of Manchester, the fault zone is again very wide. It is marked by parallel ridges running southeast-northwest and separated by low sag areas. Side roads out of Manchester lead to this section of the fault. The 1906 earthquake shook this area severely, and fault movement offset most of the fences, knocked down farm buildings, and tore surface ruptures through the pastures. The shallow hollows within the fault zone mark the lines of recent activity, and the eroded ridges between were once sharp wedges forced up between the breaks by the fault movements.

ALDER CREEK

The mouth of Alder Creek marks the point where the 1906 earthquake fissure left the land to dive beneath the waters of the Pacific. The bridge across the creek was destroyed by the quake, was later rebuilt, and was finally abandoned when Highway 1 was re-routed to its present location.

You can still see the old bridge site by following the old highway route when it leaves the present coast route about 1.5 miles north of Manchester. At the end of the paved road, you can look down into the creek bed at one of the supporting pillars of the bridge (not the structure destroyed in 1906, but a later replacement). The 1906 rupture generally followed the creek bed, passed within a few feet of the south bank at the bridge site, and then cut through the bluff north of the river mouth.

North of Point Arena, the fault follows an underwater course close to the coastline. The straight line and steep cliffs of the shoreline lead geologists to estimate that the main fault stays close in for some distance, causing sharp breaks along the face of the headlands. The 1906 earthquake caused considerable damage to the sparse settlements along here, lending further support to the idea that the fault is not very far out to sea.

Repeated fault action, followed by erosion, has created a straight line valley through hills north of Gualala.

A redwood tree that was split by the 1906 earthquake rupture still stands just north of Plantation Road.

Exploring the Other Major Faults

D OZENS OF OTHER FAULTS cut through California, some parallel and others at right angles to the trend of the San Andreas system. Some are known to be active, most others are presumed dead. But even those thought to be inactive can be full of surprises, as evidenced by the 1952 and 1971 earthquakes.

Most of the major faults are shown on the map on page 15. Those discussed here are important either because of their great size, or because they have produced significant earthquakes during historical times.

SAN FERNANDO FAULT

The base of the San Gabriel mountains is marked by a complex system of interlocking faults. The San Fernando fault zone is part of this system and extends approximately 8 or 10 miles from the middle of San Fernando eastward into Big Tujunga Wash north of Sunland.

The existence of these faults has been known for quite some time, but they are so complicated and so relatively insignificant that they appear on only the most detailed fault maps of the Los Angeles area.

The fault line in the San Fernando and Sylmar areas is not particularly noticeable to laymen, and the extensive man-made developments in the northern part of the San Fernando Valley tend to obliterate the few surface features that might normally be apparent.

Even though the northern San Fernando Valley has been less active seismically than other parts of the Los Angeles area, it was the scene of one of the most devastating—and important—earthquakes ever to hit southern California.

Earthquake of February 9, 1971—San Fernando
Magnitude: 6.6; Maximum Intensities: VIII-XI.

This was the strongest quake to strike the greater Los Angeles area in 50 years. It was not a super quake in terms of magnitude (one of equal size occurs in California about every four years), but it struck a

THERE ARE MANY MAJOR FAULTS IN CALIFORNIA. SOME HAVE BEEN SIGNIFICANT IN CREATING THE LANDSCAPE; OTHERS HAVE CAUSED DEVASTATION AND DISASTER.

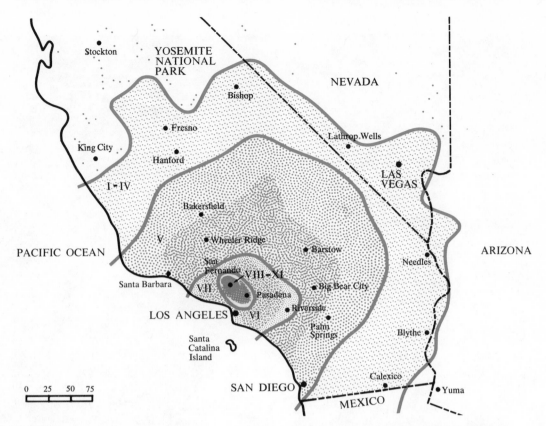

The 1971 San Fernando earthquake was felt over approximately 80,000 square miles of California, Nevada, and Arizona. The maximum intensities of VIII-XI were confined to a small area of the foothills.

heavily populated area, and the result was 65 deaths and about $500 million in property damage.

The first shock, which came at just 40 seconds past 6 a.m., had an epicenter about 10 miles north of San Fernando and 7½ miles east of Newhall (see map on page 135). The original movement was about eight miles deep, and the break continued upward and southward on a slanting plane, reaching the surface in the Sylmar-San Fernando area. A second shock of 5.4 magnitude came at 6:01 a.m. and then another four hit very close together. Ground ruptures were extensive, and in the Sylmar area, breaks occurred right in the middle of a residential area, causing the collapse of houses, rumpling of sidewalls, and extensive damage to utility lines.

In contrast to the predominantly horizontal movement associated with the San Andreas Fault system, the movement here was a combination of horizontal and vertical. The northern San Gabriel Mountain block moved in a thrust-like motion (see diagram on page 19) over the San Fernando Valley floor. As a result of the movement, the mountains seem to be higher in relation to the valley floor, and apparently have moved a few feet to the south.

In the areas of worst ground rupturing, the thrust

faulting resulted in vertical and horizontal shifts of up to six feet. The zone of rupture was about 10 miles long (from the San Diego Freeway to Big Tujunga Canyon) and the fractures were very complicated, spreading over a zone that was several hundred feet wide at some points.

The most important characteristic of the earthquake in regard to property damage was the high acceleration of the ground motion. An instrument on an abutment of the Pacoima Dam, about 5 miles south of the epicenter, registered the highest acceleration of an earthquake in the world's recorded history—in the .5 to .75 *g* range, with peaks in excess of 1 *g*. The dam itself was not significantly damaged.

This very strong motion lasted only 10 seconds, but that was long enough to cause damage greater than most seismologists would have expected from a 6.6 magnitude. As the isoseismal map on page 133 indicates, intensities were very high in the San Fernando area. Fortunately, the heaviest shaking was restricted to a small area.

The greatest loss of life was at the Veterans Hospital in Sylmar, where the old buildings collapsed, resulting in 45 deaths. The other 20 fatalities were attributed to heart attacks (9 deaths), collapse of a

This aerial photo, taken on February 10, shows the extent of the damage to the hydraulic earth-fill dam on Lower Van Norman Reservoir. A large piece of the upstream face collapsed and fell into the water. The structure still standing was badly cracked (water oozing through the cracks is visible near the top of the photo). Fortunately, the water level was several feet below normal at the time of the quake, or even the damage here would have resulted in flooding of the residential areas below the reservoir.

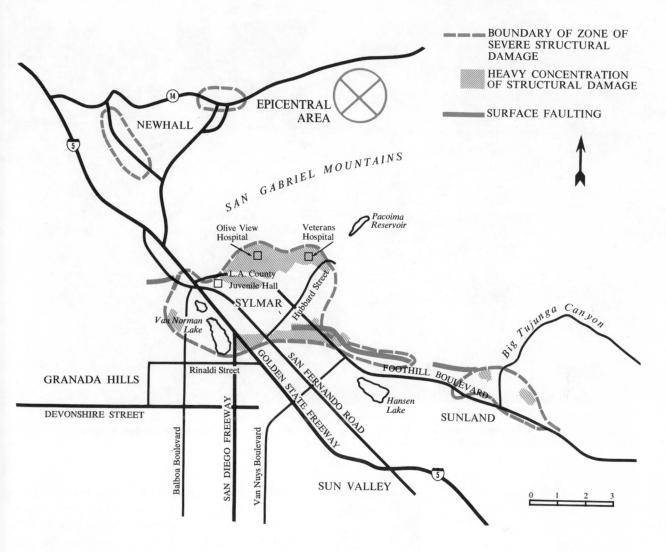

BOUNDARY OF ZONE OF SEVERE STRUCTURAL DAMAGE

HEAVY CONCENTRATION OF STRUCTURAL DAMAGE

SURFACE FAULTING

EPICENTRAL AREA

NEWHALL

SAN GABRIEL MOUNTAINS

Pacoima Reservoir

Olive View Hospital

Veterans Hospital

L.A. County Juvenile Hall

SYLMAR

Hubbard Street

Van Norman Lake

Big Tujunga Canyon

GRANADA HILLS

Rinaldi Street

FOOTHILL BOULEVARD

Hansen Lake

SUNLAND

DEVONSHIRE STREET

Balboa Boulevard

SAN DIEGO FREEWAY

GOLDEN STATE FREEWAY

SAN FERNANDO ROAD

Van Nuys Boulevard

SUN VALLEY

0 1 2 3

Map showing key areas of the San Fernando quake is based on information published by the California Division of Mines and Geology.

freeway overpass (2 dead), and other building failures in the San Fernando Valley and Los Angeles.

In addition to the Veterans Hospital, other structures that sustained major damage included the Olive View Hospital, the Sylmar electrical switching and rectifying station of Pacific Intertie, the San Fernando juvenile facility, Pacoima Luthern Hospital, Holy Cross Hospital, and Indian Hills Medical Center. Overall, it is estimated that 80 per cent of the industrial and commercial firms in the San Fernando Valley sustained some damage. The communities of Pacoima, San Fernando, and Sylmar had the worst trouble.

A potentially disastrous situation developed when the earth fill dam on Lower Van Norman Lake broke down under the intense shaking, and a large piece slid into the reservoir. But the remaining section of dam held until the water level could be lowered sufficiently to remove all danger. Some 80,000 people were evacuated from residential areas below the dam for three days and all danger was passed.

Fortunately, the water level was several feet below normal at the time of the quake. Had the reservoir been full, flooding most certainly would have resulted. Even with the lower water level, the dam came very close to failing completely. If the intense shaking had lasted another few seconds, or if a major aftershock had hit before the water was drained, failure was almost certain.

A dozen freeway overpasses in the northern part of the San Fernando Valley were severely damaged, resulting in temporary closures of portions of the Golden State, Antelope Valley, Foothill, and San Diego freeways.

SAN FERNANDO FAULT   135

Old building at Veterans Hospital had a skeleton concrete frame and unreinforced hollow tile filler walls. (Overall view of the hospital appears on page 37). Newer parts of the hospital complex built under modern structural codes performed much better than the older sections.

Los Angeles Times

Several days were required to dig out all the victims of the Veterans Hospital collapse. In general, emergency procedures after the quake worked very well.

These freeway structures had been built according to accepted engineering requirements and construction techniques, but evidently were unable to withstand the unusually high local intensities of the shaking and the randomness of earth movements.

The quake epicenter was about 30 miles from downtown Los Angeles, but 675 buildings in the city were reported to have major structural damage, and another 900 sustained moderate damage. Los Angeles' oldest building, the Villa Adobe on Olvera Street, was among the hardest hit.

It is unclear just what effect this quake had on the San Andreas fault system. No movements were recorded on the San Andreas within the first few months after the San Fernando fault break, but certainly the stresses and strains were redistributed. There is no way of knowing whether the change increased or decreased the probability for the major shift along the San Andreas that has been anticipated for several years.

However, scientists at the California Institute of Technology in Pasadena believe that the San Andreas Fault must bear partial blame for the 1971 quake and that which occurred along the White Wolf Fault in 1952. Both the San Fernando and the White Wolf

At Olive View Hospital, three people died when newly constructed reinforced concrete buildings were seriously damaged. The two story building at the top of the photo pancaked, with the second story falling to ground level. Four multi-story stairwell wings pulled away from the main building, and three (shown by arrows) toppled over.

R. Kachadoorian   Los Angeles Times

Buckling of sidewalks, curbing, and turf at Olive View Hospital was caused by subsidence of roof on parking basement below. Collapsed stairwell is in background.

Parking shelter at Olive View had a reinforced concrete roof slab supported on columns (no walls). Several vehicles were pinned when columns failed.

Among the damaged freeway structures were these at the interchange of Highways 5 and 210. Designers and engineers were unpleasantly surprisd by the heavy damage sustained by overpasses and bridges.

As rubble was being removed from collapsed section of Golden State Freeway, first teams of grim engineers began investigating causes of unexpected failures.

faults are close to the great westward bend in the San Andreas system, which runs in a straight line both north and south of this area. The bend seems to get in the way of the normal northerly movement of the western side in relation to the eastern side. Since the movement is jammed, compression builds up in the nearby transverse mountain ranges, and quakes are triggered on east-west trending fault such as the San Fernando and White Wolf zones.

The 1971 quake was very similar to the Arvin-Tehachapi quake of 1952—same type of supposedly inactive fault line, same type of thrust faulting, and even a striking comparison in aftershocks. Thirty-one days after the initial quake in 1952, Bakersfield was hit with a major aftershock (5.8 magnitude) that did more damage in that town than the original quake (see page 142).

In the case of San Fernando, a large aftershock (4.6 magnitude) came on March 31 and was centered in Granada Hills.

Generally, the aftershock pattern associated with the 1971 quake was quite normal, both in number and strength. But because they took place in a heavily populated region, they generated a great deal of publicity. There were 68 aftershocks in the first three

Dust clouds rise over the hills just after an aftershock on February 10. Veterans Hospital is in the middle distance. In the foreground is Pacoima Dam which was not seriously damaged.

hours after the main quake, and about 200 of 3.0 magnitude or greater through March 1. Almost all were north of the main line of rupture.

As with virtually all California quakes, there was a great deal of landsliding—more than 1,000 in the foothills to the east and the San Gabriel mountains to the north. However, the sliding was not an important cause of damage.

And typically, there were a number of reportings of loud noises, odd animal behavior, and strange manifestations of the earthquake force. One of the most interesting of these stories came from the Los Angeles County Fire Station at 12587 North Dexter Road in San Fernando. A 20-ton firetruck was parked inside the station, in gear and with the brakes set, at the time of the quake. When the shaking was over, firemen found that the truck had moved 6 to 8 feet without leaving any visible skid marks on the floor. In addition, there were marks on a door frame three feet above the floor that apparently were caused when the truck hit the frame. If this evidence was interpreted correctly, then the ground motion probably passed 1 *g* at this particular point.

Certainly the greatest public attention was focused on the loss of life and heavy property damage caused

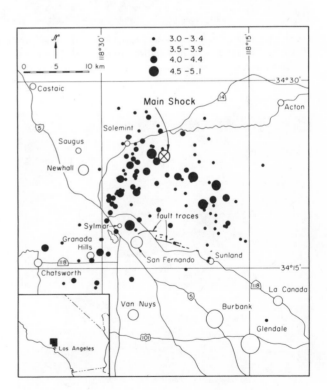

Seismological map prepared by California Institute of Technology in Pasadena shows the epicenters of the main shock and aftershocks of 3.0 magnitude or greater.

On the Montero Street cul-de-sac, scarp broke through paved street and adjacent field. Sidewalk buckled into tent shape and overthrusting was evident in the curb (see photo below). This was the first earthquake in California where a fault ruptured through a heavily developed residential area.

Robert E. Wallace

Distortions in the sidewalk on Montero Street were due to a bowing out of the curbing during thrust faulting (for fault diagram, see page 19).

by the San Fernando quake. But it was also significant in terms of the new information it contributed to geologists and seismologists. The shaking was thoroughly monitored and recorded, and our knowledge of fault movements and quake motions was greatly enhanced.

And despite the tragedies, there was a great deal of luck associated with this quake. A few more seconds of heavy shaking and the Van Norman Dam almost certainly would have collapsed. If the quake had hit at 9 A.M. instead of 6 A.M., the hospitals and public buildings would have been much more crowded and the number of deaths would have been in the hundreds. No serious landslides occurred in populated areas, and the area of intensive shaking was very limited so the emergency services of adjacent communities were undamaged and available for immediate use. Had one or two of these circumstances been altered, the 1971 San Fernando quake might have become the costliest in southern California history.

Emergency procedures generally worked very well during the first harrowing days after the quake. Communications were reliable, and evacuations were orderly. This family spent a night at Granada Hills High School, one of five emergency centers set up by the Red Cross in the San Fernando Valley.

Southern Pacific railroad tracks near the San Fernando Juvenile Hall were laterally displaced about 4½ feet in the area of maximum slippage.

Librarians all over the Los Angeles area were faced with scenes like this on February 9. Weeks of restacking, recataloging, and repair were required.

WHITE WOLF FAULT

The White Wolf Fault is unique in California. It runs between Highways 58 and U.S. 99 just below Arvin, and its known length is only 34 miles. It was first traced in the early 1900's, but was always presumed dead. Then in 1952 it suddenly and unpredictably caused the Arvin-Tehachapi earthquake, one of the largest in California history.

The fault line is almost invisible, since it is not marked by the prominent scarps and rift valleys of other major faults. In general, it lies under Wheeler Ridge, then extends northeast into Sycamore Canyon and along the northern slopes of Bear Mountain, and dies out near Caliente.

Earthquake of July 21, 1952—Tehachapi, Bakersfield Magnitude: 7.7. Maximum intensities: X-XI.

This was the largest earthquake in California since 1906, and the largest in Southern California since 1857. It was felt over an area of some 160,000 square miles and awakened people (4:52 A.M.) throughout the southern part of the state. The surface of the earth was ruptured for 17 miles between Arvin and Caliente. There were only twelve deaths; ten were in Tehachapi. The general low-population density of the area and the early hour of the shock are responsible for this surprisingly small total.

The maximum intensity was confined to a small area southeast of Bealville. The towns of Tehachapi and Arvin suffered heavy structural damage, but it was mostly confined to older brick and adobe buildings that were not adequately reinforced. Most wood-frame buildings in good structural condition withstood the shocks with very little damage, regardless of the type of rock foundation on which they stood.

Earth cracks and landslides occurred over a considerable area, as did changes in the flow of creeks, wells and springs. Caliente Creek, for example, was completely dry at the town of Caliente before the earthquake; a few days after the shock, flow was 25 cubic feet per second.

Where the fault crosses the Southern Pacific Railroad near Bealville, four tunnels were destroyed and rails were twisted and buckled. A crustal shortening of 10 feet was measured in one area.

Agricultural damage was very severe, particularly in the Arvin-Wheeler Ridge areas. Damage from the failure of irrigation systems (which was due mainly to toppled transformers for pumps) and the resultant loss of crops ran into many millions of dollars.

Several hundred aftershocks were recorded over a period of months; more than twenty had a magnitude of five or greater. The most severe of these occurred on August 22; its magnitude was only 5.8, but the epicenter was close to Bakersfield and the intensities in that town were greater than during the July 21 shock. In addition, this aftershock struck an area of downtown Bakersfield that had been substantially weakened by earlier quakes. The result was two more deaths and millions of dollars of property damage within the city.

Reports from Bakersfield residents clearly placed the August 22 quake as greater in intensity than any earlier shocks:

"The effects in my home . . . were ten times greater than any of the previous shocks. Moved a Frigidaire several inches and splashed water out of toilet tank. Window blinds flew up; all bottles overturned."

"This shock gave no warning—very sudden. No rumbling or rolling movement first as in other shocks. Much stronger than the July shocks. Broke items in house that were not disturbed previously."

G. I. Smith

This freakish situation apparently resulted when the wall of the railroad tunnel was lifted just as the track bent, then dropped on top of the rail.

The 1952 Kern County earthquake presented the first opportunity to observe the stability of earthquake-resistant construction that had been introduced in Southern California mainly as a result of the 1933 Long Beach failures. This scene of the main street in Tehachapi shows the general pattern that prevailed. The older, unreinforced buildings that characterized pre-1933 construction were badly damaged. However, modern buildings such as the two-story reinforced concrete structure in the background held up very well.

The outside walls of this hotel collapsed, but the threat to life was minimized when the interior wood partitions held firm and prevented the roof from collapsing.

This car was parked at the curb along Tehachapi's main street at the time of the earthquake. A collapsing store front landed squarely on top of it.

Clean-up crews start to clear the debris off the main street of Compton after the 1933 earthquake that originated on the Newport-Inglewood Fault. Compton and Long Beach were the hardest hit areas, but damage extended throughout the Los Angeles Basin. Cars shown above were crushed when building fronts collapsed into the streets.

NEWPORT-INGLEWOOD FAULT

This fault, which is also parallel to the San Andreas system in Southern California, lies partially under the Pacific Ocean. The trace on land starts near Newport Beach and extends northward along the Pacific coast line, past Signal Hill and Baldwin Hills to a point somewhere near Culver City. Most surface details along the fault are covered by the intense commercial development of the area, but one conspicuous escarpment is visible just east of Hollywood Park Racetrack (almost at the corner of Crenshaw and Century Boulevards).

Earthquake of March 10, 1933—Long Beach
Magnitude: 6.3. Maximum intensity: VII-VIII.

This was not a very strong earthquake, but because it occurred in a very highly developed area of southern California where heavy concentrations of commercial buildings and private residences were located on poor

Seven people were killed when this Compton building collapsed. Roof and second story fell straight down instead of toppling out into the street.

ground, it turned out to be one of the major earthquake disasters in the state's history.

In fact, this was southern California's most talked about quake until the 1971 San Fernando shake, which outdid Long Beach in terms of property damage and also relegated it to secondary status in the minds of residents.

But despite its recent demotion, the 1933 quake still rates attention for the lessons it taught Californians.

There were 120 fatalities, and damage exceeded $50 million, all of which was far out of proportion to the size of the earthquake.

The epicenter was offshore, 3 miles southwest of Newport Beach. The greatest damage was in the coastal cities—particularly Long Beach—where many unsuitable buildings had been constructed on "made"

land or water-soaked alluvium and sand.

This earthquake was significant not only for its extent of destruction and great loss of life, but also because it focused attention on the inadequacy of some types of construction—particularly that used in many school buildings. The main shock came at 5:54 P.M. Had it occurred three or four hours earlier when the schools were filled, the casualty list would have been considerably longer. A great public clamor was raised for improvements, and the state legislature passed the Field Act, which provided for improvements in building codes (see page 40).

The earthquake also presented a classic example of panic and the spread of rumor during and immediately after a major shock. One popular rumor was that the Catalina Channel had sunk 369 feet and that there had been 30 feet of horizontal movement along the fault line. In reality, there was no movement in either case.

The effects of the earthquake were bad enough without distortion. "It seemed that the highway was coming toward me in waves and the automobile became unmanageable," one man stated. "Other cars zigzagged in the road. Tall ornamental light standards along Anaheim began breaking off and showering the car with debris. I continued along toward Long Beach and had almost reached a tank farm when a series of gas tanks exploded. A transformer station went out at almost the same time, with a dazzling pyrotechnic display."

Another report came from a man standing "about twenty feet from the Balboa Island Bridge. The shock was preceded by a noise like a broadside from a battleship . . . The expansion joints separating sounded like rifle shots. I turned around and looked toward the bluffs and saw tons of dirt sliding down the road."

There were many reports of nausea and sea sickness caused by the peculiar wave motions. People momentarily lost control of their senses and often could not remember what they had done.

"I was in the kitchen on the second floor of the house, and found myself outside, but I don't know if I jumped or went downstairs."

"I found electric light bulbs in my hand."

"My mind was a blank; I came to outside."

This great calamity served to end forever the unfounded claims that the Los Angeles Basin is free of danger from major earthquakes, and to point out that great fault movements can be expected on many faults other than the well-publicized San Andreas.

Earthquake tremors caused the brick walls and supporting columns of the Young Hotel to collapse, and the second story and roof soon followed. No one was killed, despite the tangled mass of ruins.

Unstable soil conditions, particularly along the coastal areas that were predominantly fill and "made" ground, contributed substantially to the 1933 earthquake damage.

SANTA YNEZ AND RELATED FAULTS

In the southwestern corner of Santa Barbara County, a group of related faults makes up a very active earthquake region. The largest fault of the area is the Santa Ynez, but there are also the Mesa and Channel Island systems that together have caused a number of major and minor earthquakes during recorded history. But because of the complexity of the general fault area, it is often difficult to assign even a major earthquake to any one of the individual fault lines.

The Santa Ynez Fault is traceable from Gaviota Pass east along the northern base of the high Santa Ynez Mountain Range for 65 miles. It is marked by a wide zone of crushed rock and can be followed for most of its route along the base of the steep Santa Ynez mountain slopes as a sharp depression between this mountain front and the lower hills to the north.

West of Gaviota Pass, the Santa Ynez Fault branches into many offshoots but is generally traceable for another 15 miles.

Earthquake of December 21, 1812—Santa Barbara
Maximum intensities: VIII-IX.

This earthquake apparently was centered on a submarine fault between Santa Barbara and Gaviota, but it is impossible to pick out any single fault line. Santa Barbara Mission was completely destroyed, and Purisima and Santa Ynez missions sustained considerable damage. A huge ocean wave, thrown up by the earthquake movement, broke along the Santa Barbara coast. It reportedly caught a ship at Refugio, carried it up a canyon along the shore, and then returned it to sea on the backwash.

No lives were lost at the missions during this quake. However, an earlier earthquake on December 8 along an offshore fault line farther south killed 28 Indians at San Juan Capistrano. Even though these two shocks are not related, the fatality count is often attributed to the December 21 quake.

Earthquake of July 27, 1902—Los Alamos
Intensities: VIII-IX.

This quake, plus the aftershocks that lasted until July 31, caused a hectic week at Los Alamos during which the citizens became frightened almost beyond endurance. When the large shakes began on July 31, everyone left town. A special train of fourteen cars was sent from San Luis Obispo to help with the evacuation. During the initial earthquake, cracks and fissures broke the streets of the town. Not a chimney was left standing nor did a single house escape damage. Oil tanks were thrown down at Lompoc, and surface oil pipes were severely twisted and broken.

Again, it is impossible to locate this earthquake on any particular fault line. One possible source was a Santa Ynez extension north of the Santa Ynez River.

Earthquake of June 29, 1925—Santa Barbara
Magnitude: 6.3. Maximum intensities: VIII-IX.

This earthquake was centered either on a marine extension of the Mesa Fault or on the Santa Ynez system. Almost the entire business district of Santa Barbara—including all buildings constructed on loose fill—was destroyed, and the nearby cities of Goleta and Naples were hard hit. About twenty deaths were recorded. Even though there was no evident ground movement, the dam at the large Sheffield Reservoir on the Santa Ynez Fault line north of the city was destroyed.

Total property damage, not including residences, has been estimated at more than $6 million. Many of Santa Barbara's largest commercial buildings were seriously damaged, along with most public buildings, schools, churches, the courthouse, jail and public library.

This earthquake damage awakened public concern over the quality of structures being erected in this seismic area and started the reform movement that was to reach a peak after the 1933 Long Beach earthquake.

Santa Barbara and Goleta residents reported that just before the shock there was a heavy rumbling sound, similar to thunder, that seemed to come from the ground. The main shake was followed by many more aftershocks than normally expected from an earthquake of this size.

Earthquake of November 4, 1927—Lompoc
Magnitude: 7.5. Maximum intensity: X.

This was a considerably stronger shock than that of 1925, but the epicenter was on the Santa Ynez submarine extension of the Santa Ynez Fault west of Point Arguello, and the areas of heavy shaking were outside population centers. Chimneys fell and brick buildings were damaged at Lompoc, the closest town to the epicenter, and highest inland intensity was along the coast between Surf and Arlight.

A sea wave with a 6-foot rise was generated by the earth movement, and traces of it were recorded at San Francisco and San Diego.

Old church building was heavily damaged because of generally weak construction. Walls were of adobe, covered by a thin brick veneer that was not anchored to the basic structure.

Damage to the El Camino Real Garage and Hotel occurred when some of the center brick columns were knocked down, and a long girder supporting the second floor collapsed. Front of the building crashed into the street.

THE HAYWARD FAULT

The Hayward Fault is a large and active branch of the San Andreas system. It is important both as a structural element in the geology of the San Francisco Bay Area and as the center of many earthquakes, including one of the largest ever to hit Northern California.

The fault is most easily recognized as the eastern margin of the lowlands surrounding San Francisco Bay. It branches away from the main San Andreas Fault just south of Hollister, and is then traceable northward for about a hundred miles along the steep mountain front that includes the Diablo Range and the Berkeley Hills.

The actual junction of the San Andreas and Hayward Faults is hidden beneath the San Benito River flood plain. The Hayward first becomes visible north of Hollister, where a distinct escarpment juts up from the flat agricultural lands just east of Bolsa Road, 2.5 miles north of town.

Between Bolsa Road and Pacheco Pass Highway, the fault trace is marked by a straight row of small sag ponds that are located in a shallow fault valley. San Felipe Lake also lies on the fault line, and exists primarily because of ground water that has been impounded east of the fault zone.

Between San Felipe Lake and the San Jose area, the fault lies between the lowest, most westerly ridge of the Diablo Range and the main mountain ridge to the east. Coyote and Leroy Anderson reservoirs occupy long, narrow valleys within the fault zone.

Between Irvington and Niles, the fault forms an underground water barrier that has maintained the water table northeast of the fault at a considerably higher level than that immediately southwest of it. Stiver's Pond and other small lagoons nearby are within the fault zone.

Features of the Hayward Fault are most easily seen north of Niles. The line of faulting coincides with the base of the hills that rise abruptly from the valley floor along a straight line. Several offset stream courses can be seen east of State Highway 238 between Niles and Hayward, and there are also a number of landslide scars along the route.

In Oakland, the Warren Freeway lies right along the valley of the Hayward Fault, and Lake Temescal is also located within the zone.

North of Oakland, the Berkeley Hills form a very even, straight front and undoubtedly represent an ancient, eroded scarp of the Hayward Fault. On the campus of the University of California, the fault is just under the western rim of the football stadium, and there is some evidence to indicate that the most recent trace within the zone is near Founder's Rock.

In the El Cerrito-Richmond area, the line of faulting is marked by a series of shallow depressions from El Cerrito Creek Canyon to the Mira Vista Golf Course. Part of the course lies in a small fault valley.

The Hayward Fault is traceable as far as the southern shore of San Pablo Bay, where it is believed to disappear under the water just west of Pinole Point. There are a few recognizable faults north of the Bay, but none can be definitely selected as a logical extension of the Hayward.

Earthquake of June 9 or 10, 1836—Hayward area
Maximum intensity: X.

Details on this large earthquake are scarce, but great fissures reportedly opened on the surface along the Hayward Fault, and aftershocks continued for a month. Reports of damage were received from as far away as Monterey.

Earthquake of October 21, 1868—Hayward area
Maximum intensity: X.

This was one of the largest earthquakes during California's recorded history. It caused considerable damage both in the East Bay and in San Francisco. Until 1906 this was referred to in San Francisco as "the great earthquake."

A crack opened along the Hayward Fault in a straight line from Warm Springs (near the Santa Clara County line) to the vicinity of Mills College. The crack passed through Hayward, and the ground reportedly opened from 12 to 20 inches.

The greatest damage was done at Hayward, where nearly every house was thrown off its foundations. Mission San Jose's church was wrecked, and several concrete buildings in San Leandro, including the court house and the jail, were entirely destroyed.

Damage in San Francisco was confined to areas of fill and "made" ground. Most of the city felt strong tremors, however, and there were reports of strange horizontal layers of dust and clean air alternating over the main business district.

Passengers in a ferry off Angel Island felt the shock and thought that the steamer had run aground. The shaking was very heavy at Santa Rosa (nearly all major brick buildings were damaged), Healdsburg, and Guerneville.

Before 1906, the Hayward earthquake of 1868 was referred to as "the great earthquake" in the San Francisco Bay Area. It caused a great surface rupture along the Hayward Fault and did substantial damage on both sides of the bay. In San Francisco (above) many buildings were badly wrecked, and heavy tremors were felt by most people.

Temporary braces were propped against this San Francisco Building after it threatened to collapse during the 1868 earthquake.

OWENS VALLEY FAULT

The Owens Valley fault is part of the fault system that has been responsible for the formation of the Sierra Nevada and the low valleys to the east. Owens Valley itself is one of the long, narrow blocks that have been steadily sinking while the Sierra Nevada has been tilting.

One major line of faulting, which lies just east of the Alabama Hills, caused the famous 1872 earthquake. Other important breaks of the system lie at the base of the main escarpment farther west.

Earthquake of 1790?—Owens Valley
Intensity: X?

The date and intensity of this earthquake are based on Indian reports of a very large shock in Owens Valley about 80 years prior to the 1872 earthquake. However, no details are available.

Earthquake of March 26, 1872—Owens Valley
Magnitude: At least 8.3. Maximum intensities: XI-XII.

This is generally regarded as the largest earthquake in California during recorded history. It was centered on the Owens Valley fault of the Sierra Nevada system, and broke the surface for at least 100 miles from Haiwee Reservoir south of Olancha to Big Pine. The earth movements were immense, particularly between Lone Pine and Independence—some 23 feet of vertical displacement plus right-lateral horizontal shifting that totaled 20 feet in one area. Diaz Lake, south of Lone Pine, reportedly came into existence at this time.

About 125,000 square miles were sharply affected, and the earthquake was felt from Shasta County to San Diego County and in a large part of Nevada. The shocks were destructive only in Owens Valley. However, severe shaking and minor damage were reported on the western slope of the Sierra and in the foothills from Visalia and Sonora, and shocks were moderate in the California Coast Ranges. Heavy wagons were moved by the shock waves at Camp Cady military post on the Mojave River, 140 miles from Lone Pine.

The first shock—which lasted more than a minute —did most of the damage. But there were many aftershocks, and Owens Valley shook almost continuously for several days. One report estimates 1,000 aftershocks during the first three days. Only 60 deaths were attributed to the earthquake, primarily because of the very sparse settlement of the region.

Geologist J. D. Whitney, after whom Mt. Whitney is named, visited the site two months after the earthquake. He reported that "At Lone Pine we found ourselves in the midst of a scene of ruin and disaster . . . this town contained from 250 to 300 inhabitants, living almost exclusively in adobe houses, every one of which and one of stone—the only one of that material in the town—was entirely demolished. 23 persons killed . . . four more so badly injured that they have since died.

"At Fort Independence, which was entirely built of adobes, but in a very strong and substantial manner considering the material, the destruction was almost entire, and yet, strange to say, only one man was injured."

Owens River, 60 to 80 feet wide opposite Lone Pine, was dry along part of its course for several hours after the first heavy shock. One report stated that "At the foot of the bridge southeast of Lone Pine, the disturbance in the water in the river at the time of the first shock was so severe that fish were thrown out upon the bank; and the men stopping there, who were engaged in building a boat, did not hesitate to capture them, and served them up for breakfast in the morning—a quite novel method of utilizing an earthquake."

A huge wave developed in Owens Lake. Observers first noticed the water drawn away from the shore and standing in a perpendicular wall, held by the great contrasts in currents generated in the lake. But the return was fairly gentle, so only 200 feet of new ground was covered by the waves.

Perhaps the most spectacular 1872 scarp, which still appears very fresh, can be seen within a mile of Lone Pine. Take the Whitney Portal highway west out of town. After crossing the Los Angeles aqueduct, walk (or drive on jeep roads) about a quarter-mile north to where a very obvious 23-foot east-facing scarp cuts across an alluvial deposit of a former channel of Lone Pine Creek.

A picket fence surrounds the collective grave in wh
most of the victims of the 1872 earthquake are buri
The Alabama Hills and towering peaks of the Sier
are visible in the backgroun

This scarp in Owens Valley, south of Big Pine, has been the scene of considerable movement during recent geologic times, and may even have shifted during 1872 earthquake. The scarp cuts across base of a volcanic cone (right center).

Contrary to the San Andreas Fault, horizontal movement along the Garlock Fault has been left-lateral. Above, a stream channel is cleanly offset, and a steep scarp has been thrown up on one side of the fault.

GARLOCK-BIG PINE FAULT

The Garlock Fault is the east-west equivalent of the San Andreas Fault, a dominant fault zone that has had great effect on California's landscape. Starting at its intersection with the San Andreas Fault west of Tejon Pass, the Garlock may be traced for 150 miles east to Death Valley.

The Big Pine Fault, extending 50 miles west from the San Andreas west of Tejon Pass, is considered here as an extension of the Garlock, having been offset 6 miles from the main fault line.

East of Tejon Pass, the Garlock is traceable across Castaic Lake, through the Tehachapi Mountains to near Mojave, then northeast along the southern base of the Sierra Nevada and El Paso Ranges to Randsburg. It sharply separates the Basin Ranges from the Mojave Desert. For many miles east of Randsburg, the

Garlock can be traced in the long valley north of the Granite Mountains as far as the northern end of the Avawatz Mountains south of Death Valley. Its topographic expression is very similar to that of the San Andreas, particularly along the Leach Trough. The spectacular escarpments near Garlock Station were responsible for the naming of the fault.

The Garlock once was very active, and left-lateral displacements may total as much as 40 miles. The block north of the fault has been uplifted several thousand feet, and the Tehachapi and El Paso Mountains form a high escarpment along the Mojave. Farther east, the situation is reversed, and the southern block forms the Avawatz Mountains escarpment. Like the San Andreas, the Garlock zone at times appears to be narrow where the fault cuts young rocks, but it is presumably just as wide in the older rocks beneath the surface.

What Lies Ahead?

CALIFORNIA'S NEXT GREAT EARTHQUAKE MAY TAKE PLACE WHILE YOU ARE READING THIS BOOK, OR IT MAY NOT COME DURING YOUR LIFETIME. BUT ONE THING IS SURE: IT IS DEFINITELY ON THE WAY.

THE MOST IMPORTANT POINT to be made in any consideration of future earthquakes in California is that fault activity will continue for countless centuries. And as long as there are fault movements, there will be earthquakes.

As industrial, commercial, and residential developments in California continue to increase, and spread the population centers in ever-widening circles, the significance of these inevitable earthquakes also increases. Since many of the state's active faults are located close to great population centers, every major earthquake carries with it a definite threat to both life and property. Many areas which were relatively uninhabited during the major earthquakes of the first part of this century are now industrial centers or are covered with great housing subdivisions. And if an earthquake of only 6.3 magnitude can cause $60 million in damage (Long Beach, 1933), the potential destructiveness of much larger quakes above 8.0 magnitude that may hit heavily developed areas is awesome. The surface rupture of the 1906 earthquake ran through miles of unpopulated mountain regions on the San Francisco Peninsula. A similar earthquake today would crack the ground along hundreds of residential streets, and millions of people now living within 20 or 30 miles of the fault line would be hit by tremendous shock waves.

EARTHQUAKE PREDICTION

The decade of the 1960's was very important in the field of earthquake prediction. Up to that time, scientists did not have enough information to even discuss the subject with much credibility. Several schemes were proposed—often with great publicity—but very few were based on scientific knowledge.

In the 1960's, however, several related campaigns began to generate the types of statistical information that some day may allow science to predict the probability of major earthquakes in a general region within a reasonable span of time.

The U.S. Geological Survey's National Center for Earthquake Research was established in Menlo Park, California, in 1966. Scientists at the Center are mapping faults, making extensive records of large and small quakes, and analyzing earth stresses before and after these shakes. Hopefully, this research will reveal meaningful patterns that will make prediction possible.

The California Department of Water Resources established a fault monitoring program in 1959 in conjunction with the State Water Project, and the strain measurements have been very important in establishing patterns of fault movement. Other government agencies and academic institutions also are contributing very critical data to the overall efforts to predict the time, place, and size of earthquakes.

One of the necessary ingredients in developing a system of prediction is time. Measurements and analyses must be made over fairly long periods before they can become meaningful. For example, many years of effort are needed to determine the rate of accumulation of stress along different parts of the San Andreas Fault, and the exact amount of stress needed to overcome the resistance of the rocks at any given point and cause an earthquake. It has been noted that triangulation surveys across the San Andreas Fault in Northern California have established that the area west of the fault line is moving north in relation to the east side of the fault at a rate of about two inches a year. The surface ruptures of the 1906 earthquake exhibited horizontal movement of 15 to 20 feet, which could mean the release of pressures accumulated over a 100-year period. On the basis of this, it might be predicted that we can expect another earthquake of equal magnitude in about 50 years, when the strain will again have built up to a level similar to that which must have existed in 1906.

This sort of thinking assumes even greater significance when applied to Southern California. If it is assumed that the strain build-up is the same all along the San Andreas Fault, then the southern section should also move once a century. Since the last great earthquake along the southern part of the San Andreas was in 1857, it would appear that another is overdue.

However, the reasoning behind this type of prediction does not take into consideration several important points. First, there is no way of knowing if all of the stress accumulating from the 2-inches-per-year drift is actually deforming the rocks and building up energy; it may very well be that the earth is absorbing part of the movement by some method that has no bearing on pressure build-up along fault lines. Since there is no way of knowing when great earthquakes occurred before 1906 in Northern California, the hypothetical 100-year accumulation period may not be at all accurate.

Second, some of the stress may be relieved by smaller earth movements that result in lesser earthquakes. This might apply not only to movements along the San Andreas system but also to related faults. For example, the 1952 Kern County earthquake (7.7 magnitude) was centered on the White Wolf Fault, but the movement may have had a significant effect on the stress patterns of a large area, including the San Andreas Fault region. The 1968 Borrego Mountain quake on the San Jacinto fault system triggered movement on three other faults.

The 1971 San Fernando earthquake might also have had some effect on the San Andreas system. It is now thought that the north-south San Andreas has a profound effect on movements along east-west thrust faults (see pages 132-141), so it is logical to assume that the thrust faulting plays an equally important role in the distribution of stresses along the San Andreas system.

Third, there is no way of knowing whether the measurements taken in Northern California can be used to predict movements on the southern section of the San Andreas Fault. There are survey data on Southern California that indicate continuous distortion similar to that occurring along the northern portion of the fault, but the period of observation has not been long enough to determine any definite pattern. The results accumulated during the next few decades will be invaluable in assessing earthquake danger in that part of the state.

In spite of all these pros and cons, and the dangers involved in using fragmentary knowledge to predict large-scale patterns, scientists are still willing to make some estimates about earthquake probability. Dr. Charles F. Richter has set up this order of probability as his personal "guess" (note that no mention is made of when these earthquakes might occur).

1. Repetition of the 1857 earthquake in Southern California. This takes into account the estimated 100-year cycle discussed above.

2. Repetition of the 1906 earthquake in Northern California. There have not been any earthquakes of great magnitude along the northern part of the San Andreas Fault since 1906 (the 1957 earthquake is not considered as seismologically significant), and the long quiet period may be significant.

3. An earthquake along the San Andreas Fault in the Central Coast Ranges, connecting the sectors which moved in 1857 and 1906.

4. An earthquake in Owens Valley equal to that of the 1872 shock. The doubtful Indian reports of a major quake in the 1790's, plus the great shock of 1872, may add up to a significant pattern that would